DOT-COM DESIGN

CRITICAL CULTURAL COMMUNICATION
General Editors: Jonathan Gray, Aswin Punathambekar, Nina Huntemann
Founding Editors: Sarah Banet-Weiser and Kent A. Ono

Dangerous Curves: Latina Bodies in the Media
Isabel Molina-Guzmán

The Net Effect: Romanticism, Capitalism, and the Internet
Thomas Streeter

Our Biometric Future: Facial Recognition Technology and the Culture of Surveillance
Kelly A. Gates

Critical Rhetorics of Race
Edited by Michael G. Lacy and Kent A. Ono

Circuits of Visibility: Gender and Transnational Media Cultures
Edited by Radha S. Hegde

Commodity Activism: Cultural Resistance in Neoliberal Times
Edited by Roopali Mukherjee and Sarah Banet-Weiser

Arabs and Muslims in the Media: Race and Representation after 9/11
Evelyn Alsultany

Visualizing Atrocity: Arendt, Evil, and the Optics of Thoughtlessness
Valerie Hartouni

The Makeover: Reality Television and Reflexive Audiences
Katherine Sender

Authentic™: The Politics of Ambivalence in a Brand Culture
Sarah Banet-Weiser

Technomobility in China: Young Migrant Women and Mobile Phones
Cara Wallis

Love and Money: Queers, Class, and Cultural Production
Lisa Henderson

Cached: Decoding the Internet in Global Popular Culture
Stephanie Ricker Schulte

Black Television Travels: African American Media around the Globe
Timothy Havens

Citizenship Excess: Latino/as, Media, and the Nation
Hector Amaya

Feeling Mediated: A History of Media Technology and Emotion in America
Brenton J. Malin

Making Media Work: Cultures of Management in the Entertainment Industries
Edited by Derek Johnson, Derek Kompare, and Avi Santo

The Post-Racial Mystique: Media and Race in the Twenty-First Century
Catherine R. Squires

Sounds of Belonging: U.S. Spanish-Language Radio and Public Advocacy
Dolores Inés Casillas

Orienting Hollywood: A Century of Film Culture between Los Angeles and Bombay
Nitin Govil

Asian American Media Activism: Fighting for Cultural Citizenship
Lori Kido Lopez

Struggling for Ordinary: Media and Transgender Belonging in Everyday Life
Andre Cavalcante

Wife, Inc.: The Business of Marriage in Twenty-First-Century America
Suzanne Leonard

Dot-Com Design: The Rise of a Usable, Social, Commercial Web
Megan Sapnar Ankerson

Dot-Com Design

The Rise of a Usable, Social, Commercial Web

Megan Sapnar Ankerson

NEW YORK UNIVERSITY PRESS
New York

NEW YORK UNIVERSITY PRESS
New York
www.nyupress.org

© 2018 by New York University
All rights reserved

References to Internet websites (URLs) were accurate at the time of writing. Neither the author nor New York University Press is responsible for URLs that may have expired or changed since the manuscript was prepared.

Library of Congress Cataloging-in-Publication Data
Names: Ankerson, Megan Sapnar, author.
Title: Dot-com design : the rise of a usable, social, commercial web /
Megan Sapnar Ankerson.
Description: New York : New York University Press, [2018] |
Series: Critical cultural communication | Includes bibliographical references and index.
Identifiers: LCCN 2017054991| ISBN 978-1-4798-7272-5 (cl : alk. paper) |
ISBN 978-1-4798-9290-7 (pb : alk. paper)
Subjects: LCSH: World Wide Web—History. | Web sites—History. | Web site development industry—History. | Internet industry—History.
Classification: LCC TK5105.888 .A54 2018 | DDC 025.04209—dc23
LC record available at https://lccn.loc.gov/2017054991

New York University Press books are printed on acid-free paper, and their binding materials are chosen for strength and durability. We strive to use environmentally responsible suppliers and materials to the greatest extent possible in publishing our books.

Manufactured in the United States of America

10 9 8 7 6 5 4 3 2 1

Also available as an ebook

CONTENTS

Introduction: Web Histories and Imagined Futures   1

1. Forging a New Media Imagination (1991–1994)   25

2. Cool Quality and the Commercial Web (1994–1995)   56

3. Designing a Web of Legitimate Experts (1995–1998)   96

4. E-Commerce Euphoria and Auteurs of the New Economy (1998–2000)   121

5. Users, Usability, and User Experience (2000–2005)   160

Conclusion: Reconfiguring Web Histories   195

Acknowledgments   205

Notes   209

Bibliography   225

Index   245

About the Author   255

Introduction

Web Histories and Imagined Futures

In 2011, a video clip titled "What Is Internet, Anyway?" became an internet sensation.¹ Leaked from the NBC archives and uploaded to YouTube, it featured off-air footage of *Today Show* anchors Katie Couric, Bryant Gumbel, and Elizabeth Vargas in January 1994 struggling to understand the internet email address displayed on screen for viewers to contact the show.

"I wasn't prepared to translate that," Gumbel tells his co-hosts, "that little mark with the 'a' and then the ring around it?" Couric suggests it might be pronounced "about" or maybe "around." Still perplexed, Gum-

Figure I.1. Full-screen graphic with internet address follows a 1994 NBC news segment.

bel asks, "What is internet, anyway?" Vargas tries to explain: "Internet is, uh, that massive computer network, . . . the one that's becoming really big now." Gumbel fumbles for the words, struggling to comprehend how it works: "What do you mean? How does one . . . what do you do, write to it, like mail?" Confusion abounds. Couric expresses surprise that you do not need a phone line to operate internet. "It's like a computer billboard," Vargas offers. Gumbel pronounces the address "NBC, GE, com" with pauses instead of saying "NBC-dot-GE-dot-com." Nobody uses the definite article "the" before "internet." Finally, an off-camera crew member explains that internet is a giant computer network, a bunch of universities connected together.

Deemed "hilarious" and an "epic fail" by those who circulated the video through social media and technology blogs, it is striking today precisely because we can hardly imagine a world where the language and conceptual map of what the internet means were not yet in place. Typically, those of us who use the internet on a daily basis no longer need to first install Winsock and the Transmission Control Protocol / Internet Protocol (TCP/IP) stack, configure modem ports, or even distinguish between the internet and the World Wide Web. Twenty years after the *Today Show* hosts struggled for the language to describe this new communication medium, the internet has become remarkably ordinary. The basic workings of browsers, bookmarks, and back buttons are common sense. Understanding an email address is no different than understanding how phone numbers and mailing addresses work.

But perhaps because we do not encounter the network of the past in the same way that we find vintage *I Love Lucy* or last season's *Big Bang Theory* on cable or rent classic films years or even decades after they were made, our cultural memory of computing seems to operate on a different scale. Website "redesigns" write over and replace earlier sites, introducing new conceptual maps, visual schemes, and functionality while creating a sense of distance from the look and feel of earlier versions. Before Facebook's "timeline" metaphor, users of the social networking site left messages on one another's "walls." Acclimating to new mental models takes a little time, but once the timeline feels "natural," the wall feels old and foreign by comparison. Within a technological and economic system that depends on a dynamic of perpetual "upgrade culture,"[2] old platforms, operating systems, and software are left behind:

"no longer supported," in the language of the technology industry. They are resurrected as objects of nostalgia—classic game consoles, eight-bit graphics, Polaroid image filters—"retro" goods that repackage memories of earlier media experiences by accentuating the gap between then and now. Too often, the historical narratives we tell about the internet and web are likewise organized as a series of upgrades from a buggy past to a more stable, more social, more "user-friendly" future.

Nowhere is this more apparent than in the popular histories of the web that became dominant in the wake of "Web 2.0." When this term first gained traction in Silicon Valley, a few years after the 2000 crash of the dot-com bubble, technology journalists and pundits were embracing the idea that blogs, wikis, social networking—in other words, "user-generated content"—heralded a new, more democratic era of participatory media. *Time* magazine announced in 2006 that its "person of the year" was "you," and the cover story breathlessly hailed sites such as YouTube, Wikipedia, MySpace, and Facebook as fostering a social revolution. This collaborative, user-generated web, the story made quite clear, was a new one: "The tool that makes this possible is the World Wide Web. Not the Web that Tim Berners-Lee hacked together . . . as a way for scientists to share research. It's not even the overhyped dotcom Web of the late 1990s. The new Web is a very different thing. It's a tool for bringing together the small contributions of millions of people and making them matter. Silicon Valley consultants call it Web 2.0, as if it were a new version of some old software. But it's really a revolution."[3] Although the term "Web 2.0" fell out of favor within the tech scene around 2009 or 2010 (curiously coinciding with another economic crisis, precipitated by the collapse of the US housing bubble), "social media" continues to be the preferred way to understand what the web means today, how it is valued, and what the web is for.[4] But these early connections linking social media with this term "Web 2.0" helped perpetuate the idea that current ways of talking about the internet involve more advanced, next-generation technologies and are therefore naturally superior to earlier efforts from the 1990s. The internet scholar Matthew Allen refers to this type of history as a "discourse of versions," which functions by attempting to bring order and mastery over an anticipated technological future by claiming control of the meaning of the past.[5] Yet the "versioning" of web history also installs a set of divisive boundaries that reinforce a technological determinist mode of historical consciousness. As we supposedly moved from gaudy GeoCities home pages to clean Facebook timelines, from spectacular dot-com flops to social-media behemoths, a picture is painted of the past that posits an emblematic shift from "read-only" static web pages to "read-write" or participatory culture.

This book aims to show that much is left out when the web's early years are reduced to the retronym "Web 1.0." Such accounts neglect the complex cultural work of making digital media in the socioeconomic context of the 1990s, a moment characterized by widespread enthusiasm for the transformative potential of information technology, particularly the internet, to upend traditional institutional structures. Accusations of "irrational exuberance" persisted alongside impassioned pronouncements of a "New Economy" that claimed the old rules governing business, economics, and social relationships no longer applied in the internet age.[6] A surge of new dot-com startups—companies such as Webvan, Pets.com, and the *Globe* that conducted all their business online—achieved near billion-dollar valuations on the public stock market before trading for pennies a mere eighteen months later. Today, the wild excesses of the 1990s, the stuff of legend, are now packaged as cautionary tales of what transpires when greed, gullibility, and grossly overstated hype trump sound business decisions. Whether known for the vertiginous stock valuations of year-old internet startups or for the (now laughable) amateur visual styles of "Web 1.0"—spinning graphics, sound effects, and background wallpaper—the stories of the early commercial web most often serve to show us how far we have come and what mistakes bear not repeating. But alongside the hype, the buzzwords, and the soaring NASDAQ composite, the dot-com era was also a period of remarkable innovation and unbridled excitement for the creative potential of a new cultural form. It speaks to how the future was imagined at the close of the twentieth century. By ignoring how the web was historically imagined and visualized and how web design was organized, evaluated, and reconfigured, today's web is often framed as the gradual realization that user experience and social platforms matter.

Indeed, as an emerging new cultural industry, commercial web design involved parsing the very meaning of the web: what it was and whom it was for; how it should look, feel, and work; who was best qualified to design it; and what principles should guide these decisions and practices. As a cultural history, this book examines how discourses of "quality" design and dominant sets of rules defining the right and wrong ways to make the web (i.e., "top ten web design mistakes that every designer should avoid") cohere and change during this period of rising speculation in internet stocks. These shifting assumptions about the early web, however, did not just manifest themselves in web practice; they often became codified as industry standards and "best practice" guidelines and materialized in the production logics of web-authoring software, where they were used in turn to reproduce particular ideological meanings about the social life of the web, including how it should properly be imagined and designed and how users ought to experience it. Dot-com design cohered, in part, by legitimating certain visions of what the web could be while disciplining practices seen as out of the step with the future.

As media historians point out, when we look closely at the cultural milieu and reception of new media technologies—the phonograph, radio, stereoscope, or telegraph, for example—it becomes clear that there are always ways in which things may have unfolded differently. Without these histories, we risk assuming that current practices are simple common sense, a "natural" result of the way things ought to be, or that they evolved this way due to sheer technological innovation or free-market competition. Why *was* the web designed the way it was? To take a term from the 1990s, what does it mean for the web to be always "under construction"? How and why did assumptions about the right and wrong ways to design it evolve and change? In other words, how did we go from a physicist's dream of global hyperspace (which is how the web's primary inventor, Tim Berners-Lee, first conceived his project) to the walled gardens of proprietary algorithms that structure search engines, usable apps, friend requests, followers, and the quest for quantifiable social influence twenty-five years later?

To understand this shift involves investigating how the commercial web became *usable* and *social* and why these terms formed such an important, if contentious, vocabulary for interaction designers, marketers, and the internet industry at large by the twenty-first century. Today, "using" (described in terms of "usability," "user-generated content," "User Experience," or "UX design") "social" media (facilitated through social-network platforms such as Twitter and Facebook) is the dominant way of talking about websites and internet applications (now known

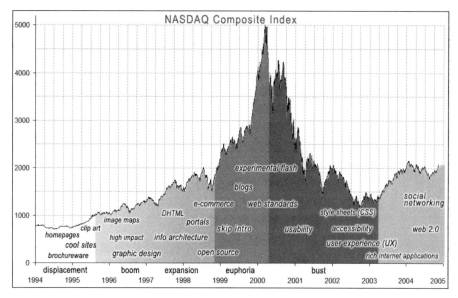

Figure I.2. Visual snapshot of the nexus of design practices and the stages of a speculative bubble mapped alongside the NASDAQ composite (1994–2005).

simply as "apps"). These terms took on particular meanings as a way to understand engagement and resurrect the internet's commercial potential in the wake of the crash of the dot-com bubble in 2000 and 2001. Indeed, this was the context in which social media and user-generated content became known as "Web 2.0." But, as we will see, "usable" and "social" were, from the very start, crucial ways of understanding what made the web distinct and valuable. The contours of these discourses, however, and how they were deployed in practice signaled quite different things at particular moments in the web's history. Attending to these distinctions, I suggest, can help dislodge the evolutionary histories that are built into the discourse of versioning. How can we reconfigure upgrade culture to organize different narratives about the past? How else might we approach web history and historiography?

What Is Dot-Com Design?

To engage these questions, I want to put the two keywords from this book's title—"dot-com" and "design"—in conversation as a way to explore the social, economic, aesthetic, technological, and industrial contexts from which a thriving commercial web industry developed alongside a growing speculative bubble in internet and technology stocks in the 1990s. Here, "dot-com" is understood as a bundle of social, rhetorical, industrial, and technological protocols (in other words, signs, rules, and conventions) that underwrote the speculative climate of the internet technology bubble. Technically, ".com" (derived from the word "commercial") is a top-level domain (TLD) in the Domain Name System (DNS) of the internet, a system that maps numerical addresses—required to be locatable online—to human-friendly textual addresses. As such, it is a technology that does particular work. If I were to title this book *Dot.Com Design* (rather than the hyphenated *Dot-Com*), such a move would actually make demands on the technical system of the internetwork: the root servers, registry, and name servers that are charged with resolving textual addresses into numerical IP addresses assigned by the internet protocol. Every time the title of a book called *Dot.Com Design* is typed into an email message, for example, the mail application will read "Dot.com" as a "URL call," a request to query the appropriate servers to achieve "resolution." It would appear in the message as a broken link, and the title could cause further confusion down the line with databases, search results, and library catalogs that are programmed to handle assets in specific ways. As a technical protocol, ".com" is code designed to translate between machines and humans.

But, of course, as a cultural and economic term, ".com" has come to stand in for much more than this. "Dot-com" signals a particular industrial configuration: the rise of internet companies in the 1990s that conducted most of their business on the internet, the perceived market opportunities and venture-capital funding that internet startups were able to attract, and the particular logics and strategies (e.g., e-commerce companies that operate at a loss to "get big fast" and capture market share) that prevailed in different moments within the dot-com bubble. At the same time, the dot-com business model cannot be understood apart from what Nigel Thrift calls "the material-rhetorical flourish" of the New Economy, in which passion for the market was also framed by the financial metrics for measuring it, thus producing and disciplining this assemblage of people, technologies, markets, and knowledge practices at the same time.[7] "Dot-com" both designated an industrial structure and actually produced a mode of valuation. Financial scholars, for example, have documented a striking positive stock-price reaction to the announcement of corporate name changes adding ".com" in 1998 and 1999, a phenomenon that was termed the "dot-com effect."[8] As a technology, industry structure, financial vehicle, and rhetoric of hype, "dot-com" recalls Foucault's notion of discourse as encompassing more than just talk: discourses have a reciprocal function as "practices that systematically form the objects of which they speak."[9]

Dot-com discourse is intensified through larger social, cultural, economic, and political forces and the everyday practices that inform how companies attract capital and workers. The financial culture of venture capital, for example, supported and reinforced particular entrepreneurial logics and expressions. This includes, as Gina Neff has argued, a style of everyday entrepreneurship that she calls "venture labor," an attitude toward risk management in which ordinary employees acted like financial investors in their jobs, investing time, energy, and personal resources in often short-term, temporary, project-based work. In this culture of high-risk, high-reward, risk became glorified, even cool, as the meaning of economic risk shifted from a collective responsibility to an individual one.[10] As the image of technological and creative work associated with the internet attracted cultural capital, "dot-com" could be understood as an attitude toward risk, as well as a lifestyle and identity. Media accounts of internet workers in New York's Silicon Alley routinely reference "dot-com kids," who were typically characterized as rich, youthful, energetic, and overconfident.[11]

More than just describing a zeitgeist of the 1990s, "dot-com" articulates an assemblage of relations into a cultural formation that is both material and semiotic. A concept that emerged from the philosophy of Gilles Deleuze and Félix Guattari, "assemblages" are groups of heterogeneous arrangements that combine and interact and, through these relations, carve a territory.[12] An assemblage is not a static collection of fixed things assembled once and for all but a dynamic process, transformative, contingent, and always in motion. It offers a way to talk about a complex of very different elements—objects, events, statements, actions, signs, technologies, sensations, bodies, passions, and so on—that enter into relation and manage to hang together for a time. Something cohered that was called "dot-com."

Just as "dot-com" is meant to evoke more than simply an internet startup, "design" is treated as an important keyword that points to much more than aesthetics, graphical layout, visual style, or the hypertext markup language (HTML) used to author web pages. Most books about web design are didactic, intended to teach the necessary technical skills, best practices, and standards that are required to create websites.[13] *Dot-Com Design* positions "design" much more broadly. Indeed, the question of what "design" even means is at the very heart of this book. Design figures as both a critical concept for assembling this historical account of the web and the form it takes in the arrangement of chapters carved along the stages of a speculative bubble. In other words, I want to invite readers to remember that this is a book about design as well as a designed book.

As a term that dates back to the sixteenth century, "design" means to mark out, contrive, plot, intend, execute. It is simultaneously a process and an object, a concept and its material expression. In this relation between the verb and noun forms, "design" collapses the distinctions between ideas and things, agency and structure, change and stability, yet nevertheless depends on carving out boundaries. By giving form and order, design "regulates": it organizes bodies, perception, affective experience, and the spatial and temporal rhythms that orient us in the world (we might think of the design of city traffic systems, home theaters, or Twitter feeds). Although translated as "assemblage" in English, the French term used by Deleuze and Guattari is *agencement*, which means "layout, put together, combine, fitting."[14] It is closely associated with design in that it retains the dual meaning of the arrangement of things and the act of arranging. John Phillips explains that unlike "assemblage," which tends to connote the finished structure of assembled elements, *agencement* prioritizes neither the parts nor the state they compose but the connections between them.[15] It is a concept that emphasizes agency as well as the mutual play of contingency and structure, organization and change. In this way, design as *agencement* offers a conceptual grip on the heterogeneous and flexible arrangements—the dense and layered connections among people, software, workplaces, proposals, contracts, skills—that enact the everyday practices of web design work. Although graphic designers or front-end interface developers are typically designated as "designers" in web design, this book intentionally positions a range of different stakeholders, from programmers to users, within the realm of design.

Since design is simultaneously a process and an object, web design histories must account not only for the actual websites and software applications produced but also for the circumstances of problem solving, the stakes and stakeholders involved, and the changing values and assumptions that inform how "design" is conceived within particular historical contexts. As the materialization of social practice, design inscribes the reigning belief systems and knowledge structures of the time. Modern design, for example, as it developed between the 1920s and the 1950s, was premised as a rational project concerned with progress and objectivity. The typographer William Addison Dwiggins (who coined the term "graphic design" in 1922) wrote a book called *Layout in Advertising* (1928), which concluded, "Modernism is not a system of design—it is a state of mind. It is a natural and wholesome reaction against an overdose of traditionalism."[16] This modern sensibility rejected ornament and decoration as forces that hindered progress, privileging instead a "neutral" and "universal" mode of communication. For the progenitors of the International Style, who had studied typography at the Bauhaus school in Germany, the aim for objective clarity corresponded to modernist social and political ideals that regarded nationalistic expressions as exclusive and unwelcome impediments to international relations and universal harmony. Grid design and simple sans serif typeforms were considered objective, clear, and impersonal, serving international communication in a way that upheld the progressive and socially improving values of modernity.[17] These values are often hidden or obscured by the very design process that naturalizes these assumptions. In tracing the drive toward functional "invisibility" that characterized midcentury modern domestic design, Lynn Spigel argues that "the attempt to make objects disappear also often winds up hiding the social relations and belief systems upon which environments are built and through which social power (in this case the uneven social relations of gender, class, and race) is organized and produced."[18]

Over the past few decades, design studies has shifted from an almost exclusive focus on the object or designed artifact to understanding design more broadly as a historically situated mode of inquiry.[19] New ways of thinking about design, particularly in the design of human-computer interaction (HCI), reject the binary divisions opposing designer/user, production/consumption, and human/machine. A key moment was the publication of Lucy Suchman's *Plans and Situated Actions* (1987), which upended the dominant problem/solution communication model that predominated in systems design at the time. Suchman was an anthropologist researcher at Xerox PARC (now PARC), and her studies of user interaction revealed that users did not approach technical systems with a discrete set of plans and goals (an assumption built into the design process) but instead improvised within the local situated context where interaction took place. Interrogating the deeply ingrained and overlapping "boundary problems" between "insiders" and "outsiders," "expert designers" and "novice end-users," that structured how design was conceived and practiced, Suchman proposed a model of design that reconfigured these boundaries by contextualizing use and incorporating the knowledge of users into the design process.[20]

Describing "the social life of design," the cultural anthropologist Arjun Appadurai argues that design is only partly a specialist activity and is better seen as a fundamental human capacity and a primary source of social order. Daily life, he suggests, is an outcome of design, a process of the everyday deployment of our energies, our resources, our ideas, and our bodies so as to accomplish results that meet our expectations.[21] Uncleaved from a purely professional context of experts, the notion of "design thinking"—figured as a user-centered, prototype-driven, iterative process of problem solving—has more recently been figured as a model for reinvigorating scholarship, rethinking public policy, and reinventing business.[22] As Lucy Kimball explains, "The main idea is that the ways professional designers problem-solve is of value to firms trying to innovate and to societies trying to make change happen."[23] Today, the concepts, language, and methods of design have expanded far beyond the fields of professional design or even business and management to infiltrate anthropology, healthcare, government, military, and public-service organizations. As Kimball puts it, design "is taking a new place on the world stage."[24]

What is it about design that it has become so culturally salient today? As companies such as Apple have made quite clear, design has high cultural value, both as a status symbol and as a way of enmeshing people, media, and devices under the banner of "lifestyle." Design's value is linked with its concern with imagining the future. For "innovators," design thinking offers a promise of creative intervention that can be successfully deployed to produce desired outcomes. In this vein, Richard Buchanan has suggested that all products of design, whether digital or analog, tangible or intangible, "are vivid arguments about how we should lead our lives."[25] Critics of "design thinking" charge that the phrase was offered as an overly neat model that packaged the future within a process format that fit within the organizational logic of the corporate world.[26] Indeed, as Suchman argues, design as a road to "innovation" prioritizes an orientation to change that "is embedded within a broader cultural imaginary that posits a world that is always lagging, always in need of being brought up to date through the intercessions of those trained to shape it: a world, in sum, in need of design."[27]

Although design is future oriented, it often materializes normative assumptions and reproduces the status quo of the time. Some designers have proposed using design as a way to counter these tendencies. Anthony Dunne coined the term "critical design" in the late 1990s as a speculative practice that could challenge the narrow "preconceptions and givens about the role products play in everyday life."[28] Invested in the ethics of design practice, critical design aims to expose the hidden agendas, assumptions, and values that are inscribed in designed things. In *Speculative Everything: Design, Fiction, and Social Dreaming* (2013), Dunne and Fiona Raby propose a critical design approach built on speculative prototyping: design is "a means of speculating about how things could be," a way to imagine alternative or more equitable futures.[29] However, it is worth remembering, as Carl DiSalvo points out, that even as speculative design tries to propose alternative futures, like any design it is nonetheless grounded in the present, and it too replicates the (often normative) assumptions, styles, and themes of the moment. Speculative design projects offer us a view of the future from the present, "reinterpreted and in relief."[30]

Dot-Com Design reimagines speculation and design by examining the era of web design in the 1990s dot-com bubble as a series of contests and collaborations to conceive the boundaries of a new digitally networked future. Throughout this book, dot-com and design are framed within a larger historical context that situates cultural production and financial markets in visual and imaginative terms that alternately uphold and transgress boundaries around masculinity and femininity, class and status mobility, knowledge and intuition, the global and the local. Struggles over expertise take place through the policing of rules in a climate sustained by the paradoxical promise that "the old rules no longer apply." In this way, the book critically attends to the ways that web design was enmeshed in systems of power that generated rules for regulating conduct and social practices; this involved right and wrong ways to think about the future, the market, the role of technology, and the construction of expertise within this conjuncture of social, economic, cultural, and industrial pressures.

Web Design Imagines the Future

Positioning design and speculation within a productive tension, this book examines how web design worked during this period as a site in which visions of the future were offered, championed, and contested. How did the ongoing struggles over aesthetics and the cultural economy of design impact the ways people imagined the web's potential? These questions are explored through two lines of inquiry: the connections between representation and collective imagination and the relationship between speculation and visuality.

First, the design of the web involved not just navigation schemes, hyperlinks, information architecture, e-commerce-powered databases, animated gifs, and elaborate Flash intros but also the production of a new imagined space, a mediated realm that, according to accounts by early web users, felt new, different, and strange and seemed to hold the potential to upend deep-rooted hierarchies. Anthropological literature that examines the concept of social imaginaries and the collective imagination can help to explore this idea more closely. As the *Today Show* clip aptly demonstrates, "getting" the internet or the web—understanding what it means, what future potential it holds, and how it brings a new kind of information space into existence—was no easy feat for those who had never encountered it before. Evaluating the context in which a shared meaning of the web first began to cohere (chapter 1) and attending to descriptions of early users' experiences of navigating the web, archiving, evaluating, and sharing links to "cool sites" (chapter 2), involves looking beyond websites themselves to the links between people and mediated representations that helped a collective imagination materialize online.

Distinguishing between the realm of fantasy and that of imagination, Appadurai suggests that "fantasy carries with it the inescapable connotation of thought divorced from projects and action."[31] It sounds private, individualistic, the stuff of daydreams. The imagination, on the other hand, "has a projective sense about it, the sense of being a prelude to some sort of expression, whether aesthetic or otherwise."[32] In this way, Appadurai portrays the imagination as a "collective social fact" realized through the formation of "solidarities," communities formed through the conditions of collective reading, criticism, and pleasure: "a group that begins to imagine and feel things together."[33] For Appadurai, the role of the global imagination was transformed in the last few decades of the twentieth century, thanks to new patterns involving the twin forces of mass migration and mass media. In this historical context, communities are capable of moving from shared imagination to collective action. Shared images, shared media landscapes, the plurality of imagined worlds become the "staging ground for action," not merely a means of escape.[34]

In this account of the early commercial web, collective imagination yokes together discourses of what cyberspace means, how it looks and feels, how it is experienced on a daily basis, and what the future portends; within this organized field of social practice, speculation meets web design. Early metaphors of cyberspace as an electronic frontier, the Wild West of information space driven by "pioneer settlers" who can tolerate the "austerity of its savage computer interfaces," were not just in the minds of a "few hardy technologists," as Mitchell Kapor and John Perry Barlow memorably described "the Net" in 1990.[35] Rather, these metaphors and ways of imagining the future of the internet were very much bound up in social realities: labor, production practices, and the social and technical protocols of computer-mediated communication and daily rituals such as visiting the "Cool Site of the Day," all worked in the service of creating a series of cultural expectations—a shared imagination—that was, in fact, quite real. Web design's social imaginary reproduces itself in the design of the web, in the conceptual models of authoring software, and in the design tutorials, link collections, and demos circulated through online forums and communities. This shared imagination is somewhat akin to the internet geek culture analyzed by Chris Kelty in

Two Bits: The Cultural Significance of Free Software. Kelty conceptualizes the set of practices connected to Free Software as a "recursive public," a commons in which geeks build, modify, and maintain the very technological conditions of their own making. What makes this possible, Kelty claims, is a shared imagination, "a shared set of ideas about how things fit together in the world."[36] Likewise, the graphical space where the user meets the web functions as an ideal arena to display the visual manifestation of hopes and dreams for the future while, at the same time, web design actively produces this future in material form.

Speculation and Visuality

Accounts of financial speculation that trigger manic episodes of madness and hysteria have enjoyed popularity since Charles Mackay's *Extraordinary Delusions and the Madness of Crowds* (1841) documented the history of Dutch tulip mania in the early seventeenth century and the South Sea Company bubble of the early eighteenth century. Today, economic sociologists and financial historians continue to examine the history of how financial markets gained legitimacy in the nineteenth century and how discourses of speculation assumed importance in popular culture at the turn of the twentieth century.[37] The deep, enduring connections between visuality and speculation, however, have received surprisingly little attention.[38]

Typically associated with the buying and selling of stocks to profit by a rise or fall in their market value, the word "speculation" is derived from the Latin verb *specere*, meaning "to look or see," and is therefore bound up with visuality and the faculty of sight. It calls up not just images of investing but lavish displays of excess and spectacular appeals to the eye. The etymology of the term also references a deeper kind of seeing that entails profound reflection on the world, a contemplation that engages a hypothetical view of the future. The economic sociologist Alex Preda notes that financial speculators in the eighteenth century were socially marginalized; they were thought to undermine the government, divert productive resources, and compromise the moral order by engaging in "radically unknowable" practices that fell outside acceptable forms of inquiry and knowledge structures of the time.[39] Financial speculation was condemned, in other words, not because it was risky but because it was incalculable. Speculation was a practice linked to moral corruption because it did not obey the laws of nature or human reason. But by the second half of the nineteenth century, a distinction was being drawn between gambling and speculation in an effort to make markets more "democratic" and hence open to middle-class family men.

A key figure in this effort was Henri Lefèvre de Châteaudun, a stock-market operator and actuary who developed a graphical representation for analyzing stock-market operations in the 1870s. Lefèvre saw financial markets as a vehicle for achieving a more just, egalitarian society. For Lefèvre, "the stock exchange was the central organ of the social body."[40] Believing that financial investments would improve the lives of the working class, attenuate class distinctions, resolve social tensions, and ensure social equality, Lefèvre (along with others, such as the Parisian broker Jules Regnault) contributed to a vernacular "science of financial investments."[41] While academic economists of the time favored abstract models as a way to try to reduce uncertainty about prices, speculators such as Lefèvre considered these causal explanations of price movements unnecessary; he set out instead to build a science of investments that was grounded in observation and calculation. He rejected abstract mathematics in favor of the concrete mathematics that structure the visible world, as used in mechanics and geometry.[42] Lefèvre proclaimed,

> The public does not need definitions and formulas; it needs images that are fixed on its mind [*esprit*], and with the help of which it can direct its actions. Images are the most powerful auxiliary of judgment; thus, whatever properties of a geometric figure result from its definition and are implicitly contained within, it would be almost impossible to extract them without the help of the eyes, that is, of images, in order to help the mind. . . . Especially in the case of stock exchange operations, where the developments are so rapid, where the decisions must sometimes be so prompt, it matters if one has in his mind clear images instead of more or less confused formulas.[43]

Here we find an account that situates speculation firmly within the realm of visuality. From seeing comes action, since "images are the most powerful auxiliary of judgment." In this case, Lefèvre's solution was to develop the first graphical representations—still in common use, according to Franck Jovanovic—that could visualize individual investor decisions in a space of coordinates.[44] By showing how options contracts could be visualized within this graphical space of horizontal and vertical axes, Lefèvre pushed for greater economic efficiency, reducing response time to market fluctuations and speeding up the flow of transactions.[45]

Over a century later, visualizing the abstract space of the market continues to be the preferred method for making financial decisions. Of course, Lefèvre's graphical instruments were a way to provide instant visibility of the outcome of complex stock-market operations; they functioned as investment instruments that guaranteed a space for financial nonspecialists to participate in the business of speculation.[46] Here, I suggest that the graphical web also played a crucial role in helping investors—both financial professionals and the speculating public—visualize the web's potential and endeavor to forecast the future. These new tools for visualizing the market in the late nineteenth century or visualizing the web in the late twentieth, however, could not accomplish the necessary work of cultural legitimation through images alone. In both cases, visual technologies were accompanied by an elaborate set of rules that aimed to discipline, transform, and rationalize the behavior of connected actors.

In the nineteenth century, the work of integrating finance into the accepted order of knowledge would require untangling the association between speculation as investment and the socially unpalatable practice of gambling. As part of this effort, literature targeting middle-class investors swelled throughout the second half of the nineteenth century: brokerage firms, newspapers, speculators, and railway engineers all published manuals, newsletters, and how-to books aimed to educate the general public on the science of investing. This slew of how-to material, Preda points out, reveals the great effort put into representing finance to the public and redefining its activities as legitimate.[47] Speculation alone does not necessarily lead to a financial bubble. In fact, speculative objects frequently appear (and disappear) in financial markets without creating a crisis or a panic.[48] Therefore, presenting financial speculation as a science of investing required developing a new set of rules for financial actions, rules that set out to rationalize financial behavior as analogous to scientific behavior: "Lack of emotions, capacity of self-control, continuous study of the markets, and monitoring of the joint-stock companies were represented as fundamental conditions of successful investments. In the first place, the notion of market behavior was stripped of its emotional, unforeseeable side, of aspects like panic, or hysteria: principles and cold blood, not passions, govern the Stock Exchange. This did not mean that financial panics did not happen or that they weren't anymore an object of reflection—quite the contrary. But panicky behavior could now be examined and explained in thoroughly rational terms as lack of self-control."[49] Feminist economists have long critiqued the construction of neutrality and the disembodied ideal speculator for universalizing masculine subjectivity.[50] As Urs Stäheli points out, this ideal speculator is constructed in gender-specific ways according to an individualistic model of complete *self-mastery*. "He lacks all emotion, and instead becomes a reflexive observer who speculates with an iron will"; the "market crowd," meanwhile, seduced into "bad" speculation, succumbs to irrational, hysterical, emotional, and volatile impulses.[51]

The cultural hierarchies embedded in discourses of legitimation are well documented and reach far beyond the economic sphere, as Pierre Bourdieu prominently points out in *Distinction: A Social Critique of the Judgement of Taste*. We can see similar discourses surface around food, clothes, cars, art, media, and literature—any cultural form where taste functions to reproduce dominant social structures. In *Legitimating Television*, Michael Newman and Elana Levine chart similar discourses in television's quest for "quality" in the era of media convergence. The cultural respectability that television earned in the twenty-first century, they argue, is built on distancing itself from "ordinary" television, which is articulated to the denigrated, feminized mass audiences of the past.[52] Similar undertones surface in the shifting logics of "good" web design, which was variously deployed to capture, condition, or contain the passions of the market. While some of these transitions can be seen as positive developments that helped institute web standards, address accessibility requirements, and prioritize *user experience*, the history of the work of disciplining web design is still fraught with troubling power imbalances.

Methods and Sources

As a cultural history of web design, this book does not tackle the applied questions that occupy web practitioners, HCI specialists, and visual-communication research consultants. I do not ask how we can best realize the full potential of the medium or ask such questions as "what is the most effective way to build a website?" or "what kind of visual practices are most aesthetically pleasing or appealing to web users?"[53] Instead, I follow the discourses that surround the shifting assumptions and provisional answers attached to these inquiries.

By cataloging and analyzing examples of web design produced in different moments of the dot-com bubble, I explore how and why dominant discourses of web aesthetics emerged, stabilized, and changed. The cultural forms, styles, and modes of production that I identify—brochureware, "cool sites," modular design, tables, grids, whitespace, print aesthetics, navigation bars, e-commerce, Flash intros, usability testing, and the like—are neither naïve attempts by early producers to create the early web nor chronological stepping stones that led the way to "better" design. Instead, I argue, these practices were the result of specific industrial conditions that served a crucial role for commercial organizations and skilled laborers testing and navigating an ill-defined territory between innovation and the familiar social norms of mediated culture.

While the work of web design in the 1990s was a truly global phenomenon, this book focuses largely on US web culture because this is where interactive advertising began and because there is little scholarship that attempts to map these early power struggles and hegemonic practices. Indeed, we need many more histories that detail web and internet production and use outside of a US/European framework. There are many different stories we could tell about the rise of the commercial web: we could track the policy initiatives of the early 1990s that led to the privatization of the internet backbone,[54] the ways that businesses responded to the commercial potential of the internet,[55] the developing industrial logics that inform data-driven marketing,[56] or the technical protocols and coding practices that enabled the internet to function as an accessible multimedia publication platform.[57] While the account presented here does address some of these topics, *Dot-Com Design* situates stories of web design alongside the awakening and reshaping of collective new media imaginations—the yoking together of images of what cyberspace means, how it looks and feels, how it is produced and experienced on a daily basis, and what new prospects it holds—to historicize the industrial and aesthetic shifts that intersect with a growing stock-market bubble.

Arguably, this book would have been very difficult to write without the Internet Archive and the WayBack Machine. One of the largest and most prominent archives of born-digital materials, the Internet Archive was founded by the computer engineer Brewster Kahle to preserve the internet's digital cultural heritage. It includes both donated digitized collections and a huge database of archived websites dating to fall 1996, assembled with the help of automated "web crawler" software programmed to roam publicly accessible web pages by following links and saving copies of files encountered as the bot completes a "snapshot" of the public web. These files are stored on massive computer drives and made accessible to the public through the WayBack Machine, a database that allows users to enter the Uniform Resource Locator (URL) of a page and view the series of copies that the crawler harvested.[58] No doubt it is a tremendous resource. However, those of us who rely on web archives to conduct historical research need to understand the peculiar nature of the web archive and the significant implications such archives pose for web historiography.

As internet researchers who study search engines and algorithms have convincingly argued, technologies that deliver our search results and offer recommendations are not neutral servants of the network but designed systems that come embedded with politics and values. Web archives, even open and publicly available resources such as the Internet Archive, are no different. Despite its ambitious tagline reminiscent of the Enlightenment encyclopedic ideal—"Universal access to all knowledge"—the Internet Archive, like all archives, has its biases and omissions.

Despite the designation of web archives as "archives," they are fundamentally different from the institutional archives that catalog donated materials into folders or box numbers, and they must be triangulated with a broad assortment of supporting materials. The research for this book draws on material from institutional archives and an extensive corpus of material I collected over a period of four years that covers the first decade of (mostly US and British) web design. This includes a decade of popular and trade-press articles, collections from *.Net* magazine, *Cre@te Online*, the annual *Communication Arts* interactive awards and judges' commentaries, screenshots and video of websites and demos shared with me by designers, prospectus reports filed with the Securities and Exchange Commission in anticipation of a company's initial public offering (IPO), numerous versions of early web-authoring software, and over a hundred production manuals and "best of the web" design annuals, many of which include CD-ROMs with screenshots, site mockups, storyboards, and video interviews. I screened about twenty hours of televised news reports from the Vanderbilt Television News Archive and consulted the Business Plan Archive at the University of Maryland's Robert H. Smith School of Business for business plans, presentations, and correspondence materials from failed dot-coms, screenings of the annual Webby Awards, and programs such as the *Computer Chronicles* and *Net Café*, which cover computing and web industries during the 1990s. I also draw on twenty-six in-depth, semistructured interviews (each lasting between thirty minutes and two hours) with practitioners who worked in web industries or created web content during the dot-com era.

Organizational Structure and Chapter Overview

To attend to the ways that industrial and aesthetic shifts intersect with economic ones, *Dot-Com Design* organizes these developments through the monetary theorist Hyman Minsky's model of a classic speculative bubble. Minsky held that the financial system under capitalism is unstable, fragile, and prone to crisis; he was known for being "particularly pessimistic, even lugubrious, in his emphasis on the fragility of the monetary system and its propensity to disaster."[59] Minsky's "financial instability hypothesis" helped inform the work of the economic historian Charles Kindleberger, who used Minsky's framework to provide a comprehensive history of financial crises, stretching back to the Dutch Tulip Bulb bubble of 1636.

In the Minsky-Kindleberger model, speculative bubbles may take on different patterns but typically share a number of common features in the development of a financial crisis. The chapters that follow are organized around five such stages—displacement, boom, expansion, euphoria, and bust—each with its own internal logic.[60] By mapping shifts in design practices and organizational structures alongside these stages, I aim both to provide a more detailed historical context and to examine the connections between market activity and cultural production. *Dot-Com Design* uses these moments of a speculative bubble as a periodization scheme and a heuristic for chapter organization. In the graph in figure I.2, I map these stages against the NASDAQ composite, the US stock index dominated by technology startups and seen as the main indicator of the dot-com economy. The graph serves as a visual snapshot of the nexus of design and cultural economies addressed in the book.

Chapters 1 and 2 cover the first stage, displacement, which begins when something changes people's expectations about the future. Chapter 1 offers a conjunctural analysis of the social, political, economic, and institutional context that paved the way for a privatized commercial internet in the early 1990s. Focusing largely on developments that took place between 1993 and 1994, this chapter examines how several ideological contradictions congealed in such a way to give shape to a shared new media imagination. It lays the groundwork for understanding the roots of dot-com speculation by looking both to crucial failures (Gopher, the information superhighway, virtual reality) and to successes (the World Wide Web, Mosaic, *Wired* magazine) and argues that these developments helped assemble a common vision of the interactive future. Chapter 2 looks to the popular discourse of "cool sites" and "cool links" that formed early evaluative criteria for talking about "quality" web experiences. Far from some hollow or subjective way of understanding the early web, I argue that "cool" served as the modality of displacement; it served as a semantic figure that gestured toward a new structure of feeling. I argue that the seemingly vague values associated with "cool"— useful, fun, and participatory—were operationalized by some of the first commercial websites seeking to make a splash online while avoiding a potential backlash from the notoriously antispam internet community.

Chapter 3 examines a series of power struggles that took place alongside the creation of a web industry during the boom stage, which began with the extraordinary attention the web commanded in the popular media after the Netscape IPO in August 1995. Focusing on the efforts of established media professionals (editors, photojournalists, magazine publishers, and graphic designers) to stake an early claim in web production, I offer a case study of two related projects that received significant attention in 1996—the world-in-a-day internet spectacular *24*

Hours in Cyberspace and the connected authoring product NetObjects Fusion—as a means to examine how particular conceptions of "social media," "user-generated content," and "real-time" were built into technologies for producing the web.

As the speculative bubble expanded from boom to euphoria stage following Federal Reserve chairman Alan Greenspan's policy response to the Asian financial crisis in 1998, the stock market soared alongside the astronomical valuations of fledgling e-commerce companies. Chapter 4 examines how the dominant discourse surrounding the web's industrial logic shifted from one of content creation to one of transactions. As the idea of the "New Economy" became articulated to e-commerce, the organizational structures of web industries were reoriented to tap into the capital and skills that were linked with these new dot-coms. The resulting wave of mergers and acquisitions within the web development sector created massive web consultancies striving to become a "one-stop shop" that could deliver all of the diverse needs that e-commerce solutions required. I analyze how creatives responded to this climate of heavy consolidation and industry restructuring by developing creative technical expertise in interactive multimedia software such as Flash. I argue that freelance designers and small boutiques used Flash to win back some of the power that had been ceded to the mega-agencies. By presenting an alternative vision of the internet that was grounded in experimentation, sensory experience, and storytelling, Flash designers challenged the transactional view of the web that was associated with e-commerce and positioned themselves as auteurs capable of delivering engaging, high-quality, cutting-edge websites.

After the crash of the stock market in April 2000, however, Flash websites came under fire for their stylistic excess. By examining discourses of reform that emerged in the wake of the dot-com bust, chapter 5 analyzes how the gendered disciplining of Flash and its visual expression in the interface prompted a thorough revision of web practice in which critiques of the "hotshot designer" and the "gratuitous" Flash site gave way to a new discourse of usability, which featured the user instead of the designer. I analyze how these critiques played out in two arenas. First, by examining the redesign of the Macromedia Flash MX authoring software and the introduction of new Flash programming environments such as Flex, I explore how the usability discourse made its way into the application itself. As the Macromedia development team worked to reposition the software, changes to the authoring environment and the surrounding tutorials, help documents, and sample code register a shift from being an animation and design tool to a platform for building user-centered, rich internet applications. Second, I examine how the tensions between usability and experience were expressed in a new approach: UX, or user-experience design, which emerged as the new dominant design paradigm of the early twenty-first century.

What was it like to go online in the 1990s? Who was paid to develop commercial content for brands and other organizations? How was the industry configured? Who decided what a "professional" website looked like, who it was for, what users ought to do online? How were "quality" websites evaluated? In short, this book is interested not just in how the web was designed but also in how it was conceived, imagined, and experienced by users and practitioners in different historical moments. At the same time, it aims to account for the contests and collaborations, the struggles for power and legitimacy, and the support or rejection of various rules defining the right and wrong ways to make the web that infused web discourse during this period of rising speculation in internet stocks. It examines how techniques for validating quality, value, skill, and expertise were also bound up with cultural hierarchies that are continually used to discipline web practices. Finally, it considers how these discourses have become codified as industry standards and "best practice" guidelines reproduced in the production logics of web-authoring software, where they are used in turn to reproduce particular ideological meanings about the social life of the web.

1

Forging a New Media Imagination (1991–1994)

It seemed that explaining the vision of the Web to people was exceedingly difficult without a Web browser in hand. People had to be able to grasp the Web in full, which meant imagining a whole world populated with Websites and browsers. They had to sense the abstract information space that the Web could bring into being. It was a lot to ask.
—Tim Berners-Lee, creator of the World Wide Web

The dot-com speculative bubble, which reached its zenith at the turn of the twenty-first century, is often attributed to the meteoric rise of the internet in the 1990s. But as the internet historian Janet Abbate points out, the commercially operated and communication-oriented internet of the 1990s emerged only after a long process of technical, organizational, and political restructuring.[1] The roots of today's internet date back to the Cold War context of the 1960s, a time when government-funded research in advanced computer and data communication technologies was highly valued and generously supported. This was when the US Defense Department's Advanced Research Projects Agency (ARPA) oversaw the development of the internet's predecessor, a computer network called ARPANET, designed to connect costly mainframes at different university research sites so that scientists could share computing resources. ARPANET grew from four connected sites to several dozen through the 1970s, transitioning in the 1980s from a defense research and military network to one managed by the National Science Foundation (NSF), a civilian agency devoted to general academic research.

In the 1980s, computer networking was exploding in the United States. The internet as we know it today is the result of millions of small networks connected to bigger hubs, which are connected to major "backbones"—fiber-optic arteries capable of moving large amounts of data quickly. With the widespread adoption of a set of protocols (Transmission Control Protocol / Internet Protocol, or TCP/IP) for linking heterogeneous computer networks together and managing traffic flow between them, ARPANET became "the internet"—a network of networks. On a technical level, the internet is an infrastructure and a set of rules that govern how computers and computer networks communicate with one another. Because it was built on the principle of "open architecture," in which all of the protocols that are necessary to connect to the network are publicly available, new technologies and applications can be built to run on top of the internetwork.

Two new applications for traversing the internet premiered in 1991. One was hailed as a breakthrough application that opened the internet to anyone. The other was a wide-area hypermedia information-retrieval initiative known as the World Wide Web. Most popular histories of the internet take the web for granted, presenting a narrative in which the web's arrival seems preordained. For those who grew up searching the web for anything and everything, it may come as a surprise to learn that it was this other system, called Gopher, which took off like wildfire in the early 1990s, while the web's inventor, Tim Berners-Lee, struggled to convince audiences of the web's potential. That today there could ever be any doubt that an alternative to the web's graphically rich, hyperlinked interface to the internet could have been taken so seriously only highlights the extent to which particular assumptions about the life of the web, its purpose and meaning, have become so ingrained and naturalized.

This chapter sets the stage for the chapters to come by mapping some of the disparate discourses that came together in the early 1990s to make the internet seem not only inevitable but a revolutionary harbinger of social, political, and economic change—one so powerful that it would be capable of convincing investors (both financial professionals and the general public) that the "old rules" of the economy and mainstream media no longer applied. It examines the roots of dot-com speculation by identifying the assembly and coherence of a new media imagination, shaped from concurrent developments that included crucial failures as well as successes and forged as a way to solidify the notion that a particular vision of the interactive future would "win." Through a discursive process that smoothed the tensions and contradictions that repeatedly made the future seem uncertain and at times incoherent, a shared image of the future was made common.

Displacement and the New Media Imagination

In economic terms, speculative bubbles start with *displacement*, which begins when something changes people's expectations about the future. Borrowing the term from the monetary theorist Hyman Minsky, the economic historian Charles Kindleberger describes displacement as an exogenous (outside) shock to the macroeconomic system that sets off a mania. "Some event increases confidence. Optimism sets in. . . . The rise is under way and may feed on itself until it constitutes a mania."[2] In the Minsky-Kindleberger model, the "outside shock" that sets the ball rolling can be any number of seemingly sudden developments, such as a fundamental policy change (such as deregulation), a financial innovation (such as derivatives), or a new invention (such as radio or the internet). During this early stage before a bubble, a few well-informed prospectors try to cash in on the vehicle of speculation and make very high returns, which naturally attracts the attention of other investors.[3] But before money is even made, displacement begins with a change in perception that ultimately creates a new object of speculation. As the web garnered more and more popular attention, many people assumed that it would fundamentally alter profit opportunities in computer technology and telecommunications sectors. But, as we will see, before this can happen, the ambiguities that point to uncertain futures must be displaced by an image that condenses these ideological antagonisms into a more stable formation, one poised to gather a sense of conviction.

To better understand how widely shared beliefs about the meaning of the "internet" or the "web" became an agreed-on thing, constructed as common sense, we might look beyond the pure economics of speculation and instead approach this historical moment as a conjunctural formation. That is, speculative bubbles are usually seen as an economic crisis, but they are intimately connected to cultural, political, and epistemological realignments as well. Stuart Hall and Doreen Massey describe a conjuncture as a "period during which the different social, political, economic, and ideological contradictions that are at work in society come together to give it a specific and distinctive shape."[4] Conjunctural analysis, then, involves examining the complex ways in which various distinct and often contradictory currents come together and momentarily fuse as a new articulation of the "interactive future" at a particular time and place. This chapter turns to the first half of the 1990s, when an image of the internet and especially the web began to coalesce into a definable form. As the following chapters reveal, however, this articulation is never settled once and for all. Throughout the 1990s (and, indeed, still today), the meaning of the web, including what it should look like, who should produce it, how it should (or should not) be designed, and what (and whom) it is for were constantly renegotiated, as different actors and institutions struggled to stake a claim in defining new media and directing its potential. The point is that this fragile configuration in which an "information revolution" became popularly understood and materialized into something "usable" was by no means natural or inevitable; rather, it represents a set of forces that come together as contradictions congeal. The "interactive future" could have been any of a series of alternative propositions; as the rise of Gopher shows, other things had to happen for the web to achieve hegemony.

To map the configuration of a new media imagination that came to be articulated as "cyberspace," I concentrate on a number of developments that took place between 1993 and 1994—the moment in which we might locate a transformation in the ways that interactive media, computers, and digital culture were being thought about and talked about. Of course, this is not to say that everything changed in a single year or two; many of these shifts were years and even decades in the making. But over the course of these two years, we can see the assembly of various strands accumulating to create, if not a rupture, then the conditions that define and delineate a new set of possibilities. Certainly, we should be mindful of what Gabrielle Hecht has called "rupture talk," a breaking point that obscures a more complex reality. As she demonstrates, rupture talk is not just talk; it can have material effects that are inscribed in sociotechnical practice, shaping expectations and staking claims to power.[5] Designing the early web was very much informed by the task of imagining the future; the discourse of rupture and revolution played a key role in the negotiation of power and claims of expertise that undergird this project. As the epigraph that open this chapter suggests, "to grasp the web in full," to understand what it means to imagine an abstract information space, was no easy feat for those who had never encountered it before. The process of "getting it" involved examining not just websites themselves but also the links between people and mediated representations that helped a collective imagination materialize online.

These media landscapes, of course, are not encountered through computers alone. The language and images surrounding these new cultural practices—going online, surfing the web, exploring cyberspace, traversing electronic frontiers, believing in digital revolutions, or building information superhighways—were increasingly covered in mainstream media, discussed with friends and co-workers, portrayed in cyberpunk fiction, documented in policy initiatives, and explained in user guides. Most people encountered these discourses well before they actually experienced the internet.

Reimagining Information Systems: From Hierarchies to Hyperlinks

So why was Gopher taken up so rapidly while the World Wide Web remained the province of a small group of early enthusiasts? Why was the web not adopted immediately? And what happened to allow the web to usurp Gopher by 1995?

Gopher was created by a team led by Mark McCahill at the University of Minnesota in response to the early-1990s quest for the ultimate Campus-Wide Information System (CWIS). At the time, universities around the United States were experimenting with ways to distribute information electronically (such as campus events, library resources, local information, gateways to external databases, and so on) to computer workstations across campus. Large committees were being formed to design CWIS environments, conferences were held, and various solutions were proposed and implemented; but the process, as McCahill and the programmer Farhad Anklesaria remember it, was tense and extremely political.[6] Everyone had different ideas about what an ideal CWIS looked like. The solution that McCahill and Anklesaria came up with was a simple menu-driven interface that allowed users to "burrow" around the network to access ("go for") files on connected servers by selecting choices from hierarchical menus in a drill-down navigation system (see figure 1.1). As McCahill explains, "Before Gopher there wasn't an easy way of having the sort of big distributed system where there were seamless pointers between stuff on one machine and another

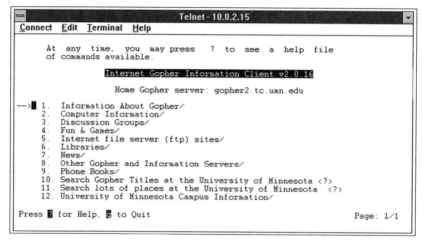

Figure 1.1. Connecting to a Gopher server via telnet.

machine. You had to know the name of this machine and if you wanted to go over here you had to know its name. Gopher takes care of all that stuff for you. So navigating around Gopher is easy."[7] Gopher (named not only for the type of navigation it offered but also for the University of Minnesota's mascot) seemed to solve the CWIS problem, so many educational institutes installed their own Gopher servers. From there, Gopher quickly proliferated.

Meanwhile, Tim Berners-Lee, a computer scientist, was working at the particle-physics lab at the European Organization for Nuclear Research, known as CERN. His idea for creating the web began in the 1980s with a desire to design a "documentation system" (software that allows documents to be stored and retrieved) that could cross-reference and link together addresses, phone numbers, projects, machines, and various types of information stored on different computers at CERN. He was frustrated with hierarchical classification systems and wanted to create a more "natural" organizational system that resembled the way a human mind can make "intuitive leaps" or random associations of thought between two ideas that at first seemed unrelated. Information, he argued, is made more valuable by what it is related to, not as it exists on its own.[8]

Although Berners-Lee put the first web page online in the summer of 1991, few people regarded this new technology as a sudden windfall;

in fact, for the first couple of years, he had to work quite hard to convince people that the World Wide Web was useful and viable. For many members of the internet community, the web's logic of linking was not intuitive. "Getting people to put data on the web often was a question of getting them to change perspective, from thinking of the user's access to it not as interaction with, say, an online library system, but as navigation through a set of virtual pages in some abstract space," he explains.[9]

The World Wide Web was growing but not nearly as fast as Gopher. In 1993, the MTV veejay Adam Curry created a Gopher site for MTV.com (later the subject of a bitter domain-name battle). Even the White House had its own Gopher, which was demonstrated on live television with much fanfare on *Good Morning America*.[10] Books such as Ed Krol's *The Whole Internet: User's Guide and Catalog* (1992), designed to introduce the internet to a general audience, made little mention of the web, focusing instead on other protocols such as file transfer protocol (FTP), telnet, wide-area information service (WAIS), electronic mail, UseNet news groups, and of course, Gopher. So what happened?

It was the combination of a few developments between 1993 and 1994 that helped the World Wide Web enter popular consciousness. According to McCahill, Gopher was really taking off, so much that "commercial guys" (such as Curry) wanted to get on board. But the rapid success of Gopher put tremendous pressure on its development team, which found that maintaining Gopher servers and developing the system for other platforms was consuming a large and growing share of resources.[11] Ongoing debates about the internet's future, especially surrounding President Clinton and Vice President Gore's promotion of a National Information Infrastructure (NII) that included plans to privatize the internet backbone (more on this later), were blurring the line between university research and commerce.[12] Responding to these debates, McCahill announced the University of Minnesota's position on the future of Gopher in a March 1993 UseNet post:

> We have been able to justify making Gopher freely available to the higher education community based on the idea that it makes more information available to us (and everyone else) on the Internet. However, when someone starts making money from our work, it makes sen[s]e for the University to get a piece of the action.... Our plan is to continue to make Gopher

freely available to the education community.... In the case of commercial use of our software we are very interested in doing licensing deals because this gets us the resources to do more development and support.[13]

Tim Berners-Lee credits the decision to charge licensing fees to commercial users as the main reason usage of Gopher fell dramatically and never recovered. Regardless of whether the University of Minnesota ever actually charged people to use Gopher, the act of reserving the right to charge commercial users set off warning bells for developers. Berners-Lee explains, "It was considered dangerous as an engineer to have even read the specification or seen any of the code, because anything that person did in the future could possibly be said to have been in some way inspired by the private Gopher technology."[14] In response to the Gopher debacle, Berners-Lee appealed to CERN to release the intellectual property rights to the web code into the general public domain, with no strings attached, not even the General Public License (GPL) developed by Richard Stallman for his Free Software Foundation. On April 30, 1993, CERN agreed to "allow anybody to use the web protocol and code free of charge, to create a server or a browser, to give it away or sell it, without any royalty or other constraint."[15]

So tensions between the commercial and public spheres were being tested before the web even garnered significant attention. For the Gopher team, it was the decision to make room for commercial uses of the technology that ultimately threatened its existence. And it was Berners-Lee's desire to make the web truly universal that prompted his decision to relinquish all rights to the code. Yet, by convincing CERN to give up intellectual property rights and put the web in the public domain, Berners-Lee helped make the web ripe for commercial exploitation in a way that Gopher could never fully realize. Unlike the Gopher team, Berners-Lee spent his time encouraging others to take up the technology—to set up their own web servers and build their own browsers. But it was not just the commercial licensing issue that led to Gopher's demise. Philip Frana argues that more than the licensing or even the "pretty pictures" for which the graphical web came to be known, it was "pretty text"—hypertext—that posed the biggest threat to Gopher.[16]

When Berners-Lee began designing his documentation system, he recognized early on that for his system to be "universal," it would need to accomplish two things. First, it had to accommodate difference, both in individual work styles and across technical environments; second, the system had to be able to scale easily so new users could join without bogging it down. Regarding the first point, Berners-Lee knew he could not require researchers to change their organizational styles and habits; numerous developers had already failed by creating systems that forced researchers to reorganize their work to fit the system. He needed to create a system with common rules acceptable to everyone running different operating systems and using different organizational models. This diversity of computing environments and organizational habits, he realized, meant grappling with the paradoxical dilemma of creating rules that were "close to no rules at all."[17] Ultimately, Berners-Lee came to regard the problem of universality across difference as a resource rather than a constraint—"something to be represented, not a problem to be eradicated"[18]—and the model he chose for his minimalist system was hypertext, a format that could display linked documents and databases in any computing environment. Presenting the idea in 1989, he explained, "We can create a common base for communication while allowing each system to maintain its individuality. That's what this proposal is about, and global hypertext is what will allow you to do it."[19]

There is a long history of "hypertext" that precedes Tim Berners-Lee. "Hypertext" was first used by the computer visionary Ted Nelson in the mid-1960s to describe "a body of written or pictorial material interconnected in such a complex way that it could not conveniently be presented or represented on paper."[20] Nelson's ambitious (and never realized) hypertext project, Xanadu, was imagined to link all the world's published information so that every quotation would have a link back to its source, allowing authors to receive micropayments each time the quotation was read. While Xanadu never came to fruition, the computer engineer Doug Engelbart, who ran the Augmented Human Intellect Research Center (now the Augmentation Research Center) at Stanford Research Institute, first demonstrated a form of working hypertext in 1968 with the "oN-Line System," or NLS. This presentation, later lionized as the "Mother of All Demos," introduced a number of experimental computer technologies, including the mouse, teleconferencing, word processing, a graphical bitmapped display (rather than a command line), and hyperlinking.[21]

Most histories of hypertext typically trace its conceptual foundations to the ideas of the engineer/scientist Vannevar Bush, who headed the Office of Scientific Research and Development during World War II.[22] In 1945, Bush published his famous essay "As We May Think" in the *Atlantic Monthly*, in which he imagines a new way of organizing and connecting information through a mechanical desk system called "the Memex." Outfitted with cameras and microfiche that would allow researchers to create linked "trails" or paths through information that could be saved and shared, the proposed technology would be a way to deal with information overload by providing an alternative to indexing as the dominant mode of information retrieval. Years before Berners-Lee, Bush's idea was a machine that could model the research process on the type of associative thinking at work in the human brain: "Our ineptitude in getting at the record is largely caused by the artificiality of systems indexing. . . . The human mind does not work that way. It operates by association. With one item in its grasp, it snaps instantly to the next that is suggested by the association of thoughts, in accordance with some intricate web of trails carried by the cells of the brain."[23] Nelson, Engelbart, and Berners-Lee have all credited Bush with inspiring their thinking about hypermedia systems. As Berners-Lee explains, "I happened to come along with time, and the right interest and inclination, after hypertext and the Internet had come of age. The task left to me was to marry them together."[24]

For Berners-Lee to meet his second requirement concerning scalability, he understood that the system would need to be completely decentralized to accommodate growth without overwhelming a central server charged with keeping track of all these relations. This meant there could be no central link database managing these connections. Links, therefore, could break. Key to the universal system was flexibility—no one link, no one type of document, could be technically more "special" than any other; all links and documents would therefore be "equal." He writes, "Hypertext would be most powerful if it could conceivably point to absolutely anything. Every node, document—whatever it was called—would be fundamentally equivalent in some way. Each would have an address by which it could be referenced. They would all exist together in the same space—the information space."[25] To consider how foreign this idea of a global information space was at the time, let us briefly consider how the internet was used and experienced in the early 1990s. In November 1993, the long-running public-access program *Computer Chronicles*, hosted by Stewart Cheifet, aired an episode titled "The Internet." One guest was Brendan Kehoe, author of the first mass-published user's guide to the internet, *Zen and the Art of the Internet: A Beginner's Guide* (1992). "What's the big deal about internet?" Cheifet asks his guest. "Why is everyone making such a fuss about it? Why is it better than CompuServe or Prodigy?" Kehoe explains that unlike commercial online services, which are directed by corporate decisions, the internet is completely user driven: "Anyone can put any service on, and have it do anything they want. . . . It's completely molded by the people who use it." When Cheifet asks him to show some of the neat things you can do on internet, Kehoe fires up a blue terminal screen and begins rapidly typing commands, narrating as he goes. "The first thing we'll be doing is looking for a job for a friend of mine who just got her teacher's certificate. . . . So we'll be using a tool called Gopher, which lets you burrow around the internet."

After Kehoe successively drills down the Gopher menu system to locate a substitute teacher position in the Northeast, viewers are taken on a whirlwind tour across various internet protocols. Kehoe next turns to one of the "more silly" parts of the internet called "finger," which allows a computer user to check if people are logged into various systems; however, he explains, some internet users have set the finger output to deliver customized information. He exits out of Gopher and returns to the command line, where he types, "% finger buckwr@rpi.edu," and indeed, this user has configured the finger output to deliver the latest top *Billboard* charts. Number five on the list is Mariah Carey, so Kehoe attempts to use telnet to access a service called Compact Disc Connection to buy a Mariah Carey CD. The system hangs, so he exits that connection and heads back to Gopher to access an editorial on the North American Free Trade Agreement (NAFTA) from the *New Republic*'s server. Finally, he uses anonymous FTP to log into another system at the University of North Carolina, where we can enter a political-science directory and find the full text of NAFTA. It is as simple as that!

If this account of "surfing the internet" seems confusing or strange to us now, it is because the difference between various networks and protocols are now masked by the illusion of a single, unified virtual information space—a space that did not exist before the web. Essentially, the World Wide Web introduced a new form of human-computer interaction (HCI) by bringing hypertext to the internet, which made every resource connected to it reachable through a common protocol. This was possible with what Berners-Lee regarded as "the most fundamental innovation of the Web": the URI (Universal Resource Identifier, later renamed Uniform Resource Locator, or URL).[26] The URI introduced an addressing scheme that, when followed, made any file or resource on the internet equally easy to link to—including files accessible through other protocols such as Gopher, FTP, or WAIS—no matter how buried the file was in subdirectories.[27] The experience of navigating through this infospace was akin to what Alan Liu calls "lateral transcendence": moving horizontally across a web-like surface, rather than digging down through directory structures.[28]

This new mode of interaction was not intuitive for many computer users at the time. Some people felt hypertext was disorienting, a "twisty maze," and they feared getting lost in hyperspace when following links.[29] It did not have the same sense of authority as other carefully ordered information systems do; library card catalogs, encyclopedias, and dictionaries are all information systems that depend on a fixed underlying structure to give them meaning. The web is decentralized and nonlinear; for many people, this felt messy and incoherent. Even the hypertext community, which was coming together in the 1980s around new stand-alone hypertext applications such as HyperCard and StorySpace, was cynical and remained unconvinced of the web's potential: "Too complicated," they reportedly told Berners-Lee.[30] Others insisted on a central link database to ensure there would be no broken links. "I was looking at a living world of hypertext, in which all the pages would be constantly changing," explains Berners-Lee. "It was a huge philosophical gap."[31]

But by around 1993 and 1994, this gap was narrowing. As Frana points out, one of the key reasons for Gopher's decline and the web's ascendency involved changing cultural assumptions about the way information ought to be organized. This did not simply relate to computers but to a broader restructuring of political, economic, and cultural sensibilities that had been unfolding over at least two decades. Hierarchies and underlying structures were losing their authority alongside a transition in the ways knowledge, representation, and "truth" were being theorized and expressed in cultural forms such as fashion, architecture, film, literature, television, and graphic design. The precise nature of these transformations involved a set of embroiled debates over what was being called "postmodernism," a movement grounded in perceived changes in the social, economic, and technical conditions of everyday life. Jean-François Lyotard famously defined postmodernism as "incredulity towards metanarratives," a rejection of the totalizing systems, hierarchies, and universal truths associated with modernism and Enlightenment progress ideals, positivist science, and the broad interpretive frameworks of Marx or Freud. Refusing grand narratives about history and humanity that serve to legitimate certain knowledges and cultural practices, Lyotard emphasized plurality, the local, and difference over claims of universality.[32] Hypertext represented a near-perfect embodiment of the postmodern challenge to sequential narrative, unitary perspectives, and the hierarchical logic of centers and margins.[33] George Landow and Paul Delany remarked in 1991 that "hypertext creates an almost embarrassingly literal embodiment of such concepts."[34] In this context, "the web epitomized the fluid and opaque postmodernist ethos," explains Frana. "Gopher, by contrast, cleaved to the classical-modernist aesthetic of technology as one-dimensional, systematic, and transparent, with 'depths that could be plumbed and understood.'"[35] As the flexibility of hypertext became a value, Gopher's fixed hierarchy became a liability. As Berners-Lee has observed, the timing was right. The idea of hypermedia now fit within this historical conjuncture in a way that it did not in the 1960s, when Nelson first coined the term "hypertext" and Engelbart demonstrated the NLS.

Thus, Berners-Lee's vision of a single, universal, global information space is simultaneously framed as a system that explodes totalizing paradigms by offering new modes of reading and writing that revel in multiplicity, open-endedness, and nonlinearity. This apparent contradiction between a master information grid and the embrace of heterogeneity and difference is deeply ingrained in the logic of the internet. Indeed, as Tiziana Terranova suggests, it is this apparent duality that enables the political potential of network culture.[36] Conceived and evolved as a network of networks, the internet is based on the premise of open architecture, driven by the problems of connecting incompatible and autonomous networks in a way that accommodates continuous expansion and modification. As we saw with Berners-Lee's efforts to create a system that would accommodate different operating systems and organizational logics, internetworking efforts are "crucially concerned with modulating the relationship between differentiation and universality."[37] In fact, Terranova argues, the challenge that network culture poses is to think both the singular and the multiple at once.[38]

Postmodernism occupied literary and cultural theorists in the late 1980s and 1990s, but there was considerable disagreement about how to characterize this "condition" at the end of the twentieth century and whether it constituted a break with the modern epoch or was instead, as Stuart Hall suggests, one emergent tendency (among others) growing out of modernism as it has developed historically.[39] For the Marxist critics Fredric Jameson and David Harvey, postmodernism is framed within the cultural logic of advanced or postindustrial capitalism. Jameson speaks of a new kind of depthlessness or "superficiality" that corresponds with transnational consumer economies based on global capitalism, while Harvey argues that postmodernism was part of a "sea change" of developments marked by a shift to a more flexible mode of accumulation (what others have called "post-Fordism" or "flexible specialization") and the compression of new forms of time-space experience resulting from rapid reduction in transportation and communication costs, just-in-time production, and the interlinking of global economies.[40]

With the fall of the Berlin Wall in 1989 and the rise of postindustrial knowledge work, networked microcomputers, globally connected stock exchanges, and the triumph of free markets through the 1990s, there was a sense that capitalism had emerged from the Cold War victorious. We might, then, situate the rise of the web in the last decade of the twentieth century as characterized by some peculiar cultural and economic tensions: on the one hand, the condition of postmodernity with its rejection of grand narratives and universal truths nevertheless fit within the "universal" model of the web as a global information space; on the other, the "incredulity of metanarratives" coincided with an unwavering faith in the logic of markets and the rise of neoliberalism, an ideology that Thomas Frank has called "market populism." This belief that markets were "a popular system, a far more democratic form of organization than (democratically elected) governments," was itself a discourse riddled in contradictions; it spoke of economic fairness, decried elitism, and deplored hierarchy at the same time that it elevated the wealthy and made the corporation one of the most powerful institutions on the planet.[41] The trust and acceptance that these two abstract information spaces—the web and the market—were able to summon were key conditions of this new social imaginary being forged.

Imagining the Future

As Berners-Lee predicted, it would take a browser to help people fully grasp the abstract information space that the web could bring into being: a window through which a user could access and "see" cyberspace. Safari, Chrome, Internet Explorer, and Firefox are all examples of contemporary browsers, applications for viewing web pages and navigating from one website to another by clicking hyperlinks. Browsers are used to retrieve files from web servers (computers connected to the internet that store and transmit files) and then format and display text and images according to instructions specified in hypertext markup language (HTML), the code that describes the structure of a web page. Without a web browser, then, there is no web to see.

In 1993, a new web browser called Mosaic was launched, created by a team of programmers led by Marc Andreessen and Eric Bina at the National Center for Supercomputer Applications (NCSA) at the University of Illinois, Urbana-Champaign (see figure 1.2). Commonly credited with jump-starting the web's popular appeal, Mosaic made the web visible and imaginable to a new audience of users. Although it was not the first web browser (and not even the first graphical browser), Mosaic ushered the web into popular consciousness for several reasons. First, unlike earlier browsers, it was easy to install.[42] Second, Mosaic introduced "in-line images," or the ability to display text and graphic files together in a single window, rather than having images open in separate windows. This meant that more than any other browser, Mosaic came closest to displaying web pages that resembled print or CD-ROMs. Because the inviting interface required less of a learning curve before a user had point-and-click access to the web, it was picked up more rapidly than the other browsers were. Finally, Andreessen's dedication to fixing bugs and responding to user requests day or night lent the Mosaic experience an air of professionalism.[43]

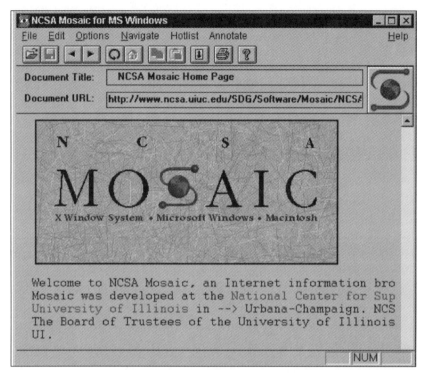

Figure 1.2. NCSA Mosaic browser for Macintosh.

Throughout 1993, Andreessen and the NCSA Mosaic team continued to release updates to the browser, which initially ran only in the X-Window environments of popular Unix systems such as Sun, IBM, and SGI workstations. But in November, the first "official" version of Mosaic was released, and by then, the browser could be installed on a much greater number of machines, including the Mac and Windows operating systems that were common on personal home computers. With Mosaic, the web truly began to take off. In April 1993, there were sixty-two web servers; by May 1994, there were 1,248.[44]

Yet, as Thomas Streeter points out, although Mosaic was a useful contribution, it was certainly no more important than were some of the other technical contributions and protocols, especially the Hypertext Transfer Protocol (HTTP), a protocol that enabled the exchange of web pages in the first place.[45] Mosaic did not enable internet connectivity, nor did it necessarily make the internet "friendly" or more efficient.

Instead, Streeter suggests that Mosaic created an instant "wow effect" and became the "killer app" of the internet—generating a flurry of interest that took the internet away from the realm of technical experts into broad popular consciousness—because it was, quite simply, pleasurable. The type of pleasure that Mosaic offered, according to Streeter, was "the pleasure of anticipation," in which the anticipation of pleasure becomes part of the pleasure itself.[46] Browsing the web with Mosaic was so pleasurable, he argues, not for what it helped users find online (there was still relatively little there) but because it inspired them to imagine what else they might see: "To engage in the dreamlike, compulsive quality of web surfing in the early days was an immersion in an endless what's next?"[47] In other words, Mosaic's contribution to this moment of displacement was not for any technical accomplishment but because, in clicking through links and watching grainy images slowly appear, Mosaic provided the first prominent site for users to imagine the future of the web.

The graphical space where the user meets the web functions as an ideal arena to display the visual manifestation of internet hype. As opposed to the hierarchy of files that users traversed through Gopher, Mosaic introduced the web as a visual representation of text, images, and hyperlinks that could be not just published but designed. Perhaps, then, it is no surprise that as the increasingly graphical web entered popular consciousness, the business of web design became such a serious industry. Mosaic gave early users their first glimpse of the web and set the stage for web design to become a prominent new cultural industry while helping users to imagine what the internet might look like in the future. As the old idiom goes, seeing is believing.

The Information Superhighway: A Crisis of Representation

Speculative bubbles begin when some new object or circumstance is perceived as a displacement, something powerful enough to alter the way investors imagine the future. As inspiring and popular as Mosaic was, we cannot fully appreciate this conjuncture merely through the success of a certain browser that made the web pleasurable. As with the web's ascendency alongside the decline of Gopher, it is crucial to consider the failure of other visions that did not coalesce. I therefore place the events already mentioned (the web's representation of a new hyperlinked, abstract information space and the pleasure of anticipation foregrounded by the Mosaic browser) in context with three other developments that helped condense various perceptions of the way the future was imagined: the unsuccessful launch of a series of interactive television trials inaugurated in the name of the "information superhighway"; the failure of virtual reality as a premier site of cyberspace; and the successful debut of the new techno-culture magazine *Wired*. While the latter offers a specific vision of digital culture as a profound force of revolutionary change, the two failures were partly responsible for relocating the hype surrounding the information superhighway from the television set to the computer screen, where the online realm of data networks became more commonly known in the popular imagination as "cyberspace."

Just as universities around the country were debating ways to wire their campuses and build CWISs, there was interest in the late 1980s in developing a high-speed national computer network. One committed champion was then-senator Al Gore Jr., who was crafting legislation to support a new National Research and Education Network (NREN), a multigigabit network available not just to researchers but to schools, libraries, and community colleges in every state. "Gore's bill," as the High-Performance Computing Act of 1991 was called, authorized $2.9 billion in financing over the following five years—a portion of which would support the development of the NREN, which was expected to replace the internet. To translate his policy initiatives to the American public, Gore used a much catchier moniker: he called it the "information superhighway."

Although by late 1994, most people used the term "information superhighway" interchangeably with "the internet,"[48] the notion of a nationwide high-capacity "superhighway" for electronic data was a much-hazier concept that did not, at first, refer exclusively to computer networks or the internet but a storied national fiber-optic internetwork of phone, cable, and computer networks in the service of education, industry, healthcare, science, and entertainment. It was never a single idea but an amalgam of different visions of the digital future.[49]

After presidential candidate Bill Clinton announced Gore as his running mate in the summer of 1992, the technology policies that Gore advocated as a senator were folded into Clinton's "Rebuild America" plan to revive the economy by restoring the country's infrastructure through a massive public-works project: rebuilding roads and bridges and constructing a national information network that would "link every home, business, lab, classroom and library by the year 2015."[50] While Gore's initial vision of the information superhighway was conceived as a public network that would be constructed and regulated by the government and operated by private industry, the initiative was renegotiated throughout 1993 as the newly elected Clinton-Gore administration attempted to calm industry concerns over the role of the private sector in building, owning, and operating the superhighway. Indeed, Paul Ceruzzi notes that Gore's 1991 vision of a future internet was nearly the complete opposite of how things actually turned out.[51] When the administration unveiled its *National Information Infrastructure: Agenda for Action* report in September 1993, the public-works language was quietly dropped. Instead, the report stated that "the private sector will lead the deployment of the NII," while the administration would promote private-sector investment, pass communication-reform legislation, and extend universal service requirements.[52]

By this time, the race to build information highways was already well under way as phone, cable, satellite, and computer companies scrambled throughout 1993 to stake their claim to the interactive future.[53] Phone companies envisioned two-way traffic such as video conferencing, database access, and "video dial tone" services (the transmission of video over telephone lines) to offer movies on demand. Cable companies imagined an entertainment universe consisting of "500 channels," video on demand, and home-shopping networking.[54] Each regarded the information superhighway as something that fell squarely within its own purview. Michael Wolff recalls, in a personal chronicle of his adventures launching an internet startup company in the early 1990s, his dealings with "old media" companies that were making the leap to new media: "Within Time Warner, throughout 1993, people would say, 'We're building the information superhighway in Orlando.' It was not for them a metaphor. They thought they were building it, the actual information superhighway thing."[55] Time Warner's planned "Full Service Network," as its information superhighway was formally called, was announced in January 1993, garnering substantial media coverage. The new pilot program would ambitiously deliver interactive services, games, home shopping, movies on demand, and telephone services to 4,000 test homes by early 1994, with a national rollout to follow; it was paraded as a "radical rethinking of what television means."[56]

The hype surrounding information superhighways swelled throughout 1993 amid news of numerous experiments with interactive television services: Viacom, TCI, GTE, and Bell Atlantic all announced their own information superhighways that would link hundreds of communities through a nationwide fiber-optic network. "Interactive" was the preeminent buzzword of the day. Larry King explained what this would mean for television viewers: "Movies that start whenever you want them to start, interactive programs you direct from your chair. Play computer games; sell stocks; buy plane tickets; order pizza—you name it. They're calling the new system a 'data superhighway,' and your TV set will be your on-ramp."[57] These dreams were not presented as futuristic; media accounts consistently reminded their audiences that new video services were not years away but months, available to millions of Americans in all fifty states by "this time next year."[58]

But the interactive tests were not under way for long before companies realized they lacked competencies in areas that were not historically part of their core business. Cable, after all, was traditionally a one-way communications business, and the superhighway would require familiarity with advanced two-way switching systems. To be the first to construct the information superhighway, a frenzy of new alliances formed across industry sectors: US West teamed up with Time Warner; Nynex invested in Viacom; Cox Cable partnered with Southwestern Bell. Dubbed "cable-phone mania," the deal making reached its zenith with the October 1993 merger announcement (the largest in US history to date, valued at up to $33 billion) between the cable provider TCI and the regional phone operator Bell Atlantic. Commenting on the storm of mergers and alliances surrounding superhighway mania, *Newsweek* observed, "It resembles the historic contests to drill oil or build railways, and if even the experts don't know what the future looks like, they are nevertheless in a mad dash to buy land rights and lay track."[59]

The regulatory climate surrounding media and telecommunication industries was undergoing vast revision as antimonopoly laws were struck down in favor of deregulation and neoliberal economic policies.

The subsequent deal making surrounding media convergence helped buoy "high-tech" or "information superhighway" stocks (cable, computer, phone, and entertainment offerings) in 1993 and 1994, creating a climate touted as a "wildly lucrative business opportunity" for which the financial-services and investment-banking sectors readied themselves. "There is a pattern, albeit a painfully simple one," reported *Investment Dealers' Digest* in December 1994. "Since no one knows what the future of media will be, the big players are placing bets all across the board."[60] For the investment-banking groups that managed these deals, the stakes were believed to be staggering, an epochal change in the ways business is done and people live their lives. The rhetoric was breathless, as a Goldman Sachs partner aptly summed up the scene at the end of 1994: "In the history of man, there has never been a bet like this."[61] A Bear Stearns report that same month advised, "In our view, the creation of a fully interactive nationwide communications network could open up *the largest market opportunity in history*, possibly generating several hundred billion dollars in new net GNP growth over the next fifteen years."[62] In other words, the big players that correctly anticipated the shape of the changing media and communications landscape and placed the winning bets on the interactive future would be in place to exploit anticipated new forms of wealth creation and lead the new media industry for decades to come; those who failed to foresee the right future would be weakened or destroyed.

The early interactive-television version of the information superhighway started to unravel in 1994. It began that February, when the "deal of the decade" between Bell Atlantic and TCI was called off, after Congress mandated cable rate reductions that reduced the value of TCI.[63] Less than a week later, Time Warner announced that the Full Service Network launch would be delayed until October, citing software and hardware delays by suppliers that were having trouble getting the technology to work. By October, only two of the projected 4,000 homes were connected. Furthermore, when other interactive-television tests were launched, viewer response was deemed "tepid."[64] These failures and disappointments inspired endless twists on the tragic highway metaphor; "potholes," "roadblocks," and "detours" all bedeviled the information driver. It was not until December 1995 that Time Warner reached its goal of 4,000 homes. By that time, the interactive future was already pinned to the web.

Visualizing the Interactive Future: Hype, Hope, and Dashed Dreams

The hype about the interactive future did not simply disappear when the cable-television/telephone superhighway failed to materialize. With the rising popularity of the web and the Mosaic browser, along with increasing mainstream-media coverage of the internet, the discourse of the interactive future was rearticulated in 1994 from the TV to the PC. With this, the family-friendly and commercially sanitized image of the information superhighway promised by cable and phone companies collided with an entirely different vision, one that painted the network as uncharted territory, a wild frontier filled with intrigue and adventure.[65] As Steven Levy summed it up in 1995, "If the dream of having 500 channels is an outdated future, the new future is internet on your computer screen."[66] The infobahn was here, just via a different "on-ramp." The dream of established media and telecommunications companies providing access to regular, orderly, organized information superhighways underwent a crisis of representation.

At first, the industry's primary role seemed practically assured. Interviewed for an *ABC World News Tonight* segment in September 1993, ABC president Bob Iger expressed a certain confidence that traditional media institutions would be necessary in the interactive future: "We're very comfortable for the consumer; we are *there*, we're there all the time, and we're there in a very regular, orderly fashion. And I believe that in a world where there is massive choice, there's still going to be a need on the part of the consumer for someone else to create an order."[67] Likewise, in the 1995 book *Road Warriors: Dreams and Nightmares along the Information Superhighway*, the financial consultant Daniel Burstein and the journalist David Kline provide a detailed analysis of the maneuverings and initiatives of the CEOs, technological visionaries, investors, and policy makers who pursued the information superhighway between 1993 and 1995. After interviewing dozens of the biggest players, Burstein and Kline confidently conclude, "It is our belief that [the internet] will *not* become the principal Information Highway that corporate America depends on for its most critical data and financial traffic."[68] In their reasoning, they attempt to envision the internet's future by identifying the core characteristics of its "personality": "Free. Egalitarian. Decentralized. Ad hoc. Open and peer-to-peer. Experimental. Autonomous. Anarchic." These qualities are contrasted with the vocabulary of business and commerce: "For profit. Hierarchical. Systematized. Planned. Proprietary. Pragmatic. Accountable. Organized and reliable."[69] The internet was seen as much too crude. Its narrow bandwidth hosts "text-only" interactions that are great for cavorting with online "pen pals," they argue, but preclude the rich multimedia potential of full-motion video and sound. The internet is a "dirt road," they remark, not a "paved superhighway." What is more, the dirt road is filled with dirty back alleys—teeming with sex, drugs, hackers, digital scofflaws, security issues, libel, misinformation, copyright infringement. It is, to Burstein and Kline, a "congested and jerry-rigged system."[70] Crucially, they note, women account for a paltry 15 percent of internet users, and it is women who make the key buying decisions; it is their needs and buying habits that shape the consumer marketplace. Unlike the potential of interactive television, the internet was just not family friendly or "corporate ready."

Contrasting this "orderly," corporate-ready version of the information superhighway with the darker, more adventurous discourse of cyberspace, Thomas Streeter argues that the latter trope ultimately succeeded precisely because it was better aligned with the sense of romantic rebellion and open-ended possibility that resonated with midlevel professionals watching the arcane world of internet communication overtake the organized, managerial mode of thought associated with the traditional corporate liberal technology policy that produced the information superhighway.[71] But the trope of cyberspace was also linked to another failure, one that played a hand in elevating the feeling of the network's untamed text-only landscape over a polished 3D visual display. The story of this rhetorical move highlights the ideological currents of a particular group of internet and computer enthusiasts, many who were active members of a popular bulletin board system (BBS), the Whole Earth 'Lectronic Link (WELL), in the 1980s. As Fred Turner argues, it was through the writings of WELL users such as Howard Rheingold and John Perry Barlow that the discourse of "virtual community" and "electronic frontier" became vital frameworks through which the American public came to understand the meaning of the internet and the web as cyberspace.[72]

The term "cyberspace" was famously coined by the science-fiction writer William Gibson, who developed the idea in his 1984 cyberpunk novel *Neuromancer*, which featured a "console cowboy" computer hacker protagonist, Case, whose special talent was cracking electronic banking systems. As Gibson describes it, cyberspace was an imagined space—"a consensual hallucination"—but it was also "a graphic representation of data" and hence was always about visualizing the imagined spaces of computing.[73] As such, it served as a compelling vision of the future, which real companies launching graphically rich commercial applications were desperately trying to appropriate. That came in the late 1980s with "virtual reality" (VR). If interactive television was one version of the original "information superhighway," VR was projected as the original "cyberspace." According to Allucquére Rosanne Stone, Gibson's fictional cyberspace "triggered a conceptual revolution among the scattered workers who had been doing virtual reality research for years."[74] By articulating "cyberspace" to VR, these researchers developed a collective social imaginary of their own that enabled them to recognize and organize themselves as a community.

In 1989, the technology company Autodesk, supplier of computer-aided design (CAD) software, began developing a virtual design system that it called the Cyberspace Developer's Kit: a new computer-user interface that would allow users to don gloves and stereoscopic display goggles to see and manipulate data in a computer-generated virtual reality. The company even tried to trademark the word "cyberspace."[75] From the mid-1980s until roughly 1992, mainstream-media reports on "cyberspace" referred almost exclusively to VR. But the former Grateful Dead lyricist, freelance journalist, and WELL networker John Perry Barlow began championing another vision of cyberspace in 1990. That year, a number of high-profile FBI crackdowns on hackers had caused some consternation in WELL forums, as members became increasingly concerned by government intrusion into what many regarded as free-speech issues. In response, Barlow and another WELL user, Mitch Kapor (founder of the successful software company Lotus), co-founded a computer-liberties organization that they named the Electronic Frontier Foundation (EFF).

Although Barlow originally championed the idea of VR as cyberspace, his interactions on the WELL inspired a new articulation of cyberspace as an untamed rural frontier, where networked forums such as the WELL constituted settlements of techno-pioneers—visionaries, outlaws, vigilantes—who willingly lived off the electronic land, sharing ideals of community and the socially transformative potential of technology similar to those that once animated the 1960s countercultural-ists.[76] The metaphor of cyberspace as an electronic frontier was effective and endorsed by Gibson himself, who quickly gravitated to the notion of territory over visual representation. In a *Computerworld* essay titled "Cyberspace '90" Gibson remarks,

> To hack, in the original sense, was not bad; to hack was to be there. Be where? Cyberspace. Not the neural-jacked fantasy purveyed in those paperbacks of mine. Rather, in the altogether more crucial version of the concept as currently championed by John Perry Barlow, Mitch Kapor and the Electronic Frontier Foundation. . . . Cyberspace, in Barlow's sense, is a territory generated by technology. As such, the "territory" itself is subject to constant growth and permutation—a cybernetic Wyoming writhing in some eerie interstice between concept and silicon. Yet this territory is certainly real because we can be rousted by the Secret Service for crimes alleged to have been committed there. . . . Fascinating as the potentials of virtual reality may be, I'm more impressed by [the] metaphor of the electronic frontier.[77]

In Barlow's hands, cyberspace emphasized the feeling of uncharted land and the ethical codes of the cowboys who traversed it, rather than the visual immersion of VR or the spotless and litter-free paved superhighways of corporate interactive television. Along with Gibson's endorsement, the 1992 publication of Bruce Sterling's *The Hacker Crackdown: Law and Disorder on the Electronic Frontier* helped reinforce the highly interactive, low-res image of cyberspace as a place of human communication and connection. The technological precursor of this model was not 3D film, cable television, or stereoscopic photography; it was the telephone. If the severe and hypermasculine conditions were not yet "corporate ready," they nonetheless were able to summon a gold-rush mentality that, once packaged appropriately, did even more to help a collective imagination cohere around the coming digital revolution—and this time, the computer would take center stage.

Get Wired: Prototyping the Digital Revolution

The feeling of the frontier territory clearly struck a chord, but it was the development of a new glossy image that helped "seeing is believing" retain its critical purchase as a shared, more mainstream mode of imagining the future. The most prominent site for visualizing an emerging digital revolution in the 1990s was the pages of *Wired* magazine, launched in January 1993 by Louis Rossetto and Jane Metcalfe. It is hard to underestimate the formative impact this magazine had on the discourse surrounding digital culture at the time. As Howard Rheingold suggests in an interview commemorating the twentieth anniversary of *Wired*, "Louis provided a frame for writing about digital technology and culture that had not existed. It's hard to see now how new and exciting that was."[78] But what was significant about *Wired*, I argue, is not just the space it opened for talking about technology and culture; *Wired* introduced a specific mode of imagining the future that called forth a new type of hero, the cyberelite, and did so in highly visual ways that blurred the distinctions between design, editorial, and advertising.

The impulse for *Wired* came out of Rossetto and Metcalfe's experiences working in Amsterdam in the late 1980s on a small, offbeat publication called *Electric Word* (one of the first magazines produced on a Mac with desktop-publishing software), which reported on advanced technologies surrounding machine translation, linguistic technology, and word processing. Produced for a translation firm looking to advertise translation software, the magazine found itself speaking to very specialized high-tech companies that were not necessarily familiar with developments in related fields. As a result, Rossetto explains, they needed to be able to speak across disciplines: "We had to make a magazine that they could get into, that they could find themselves in, find the importance of the work that they were doing. So we decided to make a magazine that didn't focus just on technology, but focused also on ideas, on people, and on companies. These were some of the ideas that ultimately became part of the spirit for the start of *Wired*."[79]

It was while working on *Electric Word* that Rossetto and Metcalfe began planning a new magazine that would bring the digital revolution they were witnessing in Europe to a mainstream US audience. "We could see it so vividly," Metcalfe explained. "In Amsterdam, Philips was the Sony of its day. They were experimenting with all these data types. It was a time of great imagination about digital media. We'd been in it since the late '80s, watching it, reporting on it, and it was accelerating."[80] The formation of this digital imagination that Rossetto and Metcalfe vividly describe gestures toward a fundamental shift in the feeling of the digital. Prior to this, "digital"—at least as represented in mainstream computer publications of this time—typically signified an attribute of equipment, described in the technical details of computer hardware and peripheral components. While other publications such as *Mondo2000* and *Whole Earth Review* were already merging technological culture and lifestyle, their low production values lent an air of "underground periodical."[81] Rossetto and Metcalfe wanted something bigger. By 1991, they were ready to execute their idea: technology would be the rock and roll of the 1990s, and *Wired* would be to the computer generation what *Rolling Stone* had been to the counterculture of the 1960s.[82]

But this vision failed to attract investors among the publishing elite in New York City. It was, by all accounts, a terrible time to start a new magazine. The United States entered a period of recession following the 1987 stock-market crash; after a brief recovery, economic malaise returned with the Gulf War of 1990 and 1991. Tightened advertising budgets took a toll on the magazine industry; less than half of the 500 titles launched in 1989 made it through the year.[83] A technology lifestyle magazine was not easy to pitch. So Rossetto and Metcalfe enlisted the help of their designer friends John Plunkett and Barbara Kuhr to create a sixteen-page visual prototype that they could take with them to San Francisco to offer a taste of *Wired*'s mission and design. In contrast to the computer trade press, *Wired* focused on people—whom it would name the "digerati"—as the "wrapper for ideas."[84]

"All the computer magazines we'd seen to date had pictures of machines or people sitting with machines," notes Kuhr. As she explained the *Wired* aesthetic to the cybergeeks who would grace its pages, "No machines. We're taking pictures of you."[85] In this way, *Wired* established an image of a new digital elite—a potent, typically white, and explicitly class-based view of information labor. The text of the first draft of the famous "Manifesto" that appeared in the prototype begins, "You, the information rich, are the most powerful people on the planet today. You and the information technology you wield are completely transforming our lives, our families, our neighborhoods, our educations, our jobs, our governments, our world."[86]

If *Wired*'s content focused on people as the wrapper for ideas, equally significant was the magazine's format, a deliberate blending of editorial with radical design and mainstream consumer advertising, that, when combined, offered a pulsating neon overwrap for a wired future, one that would be inhabited by those who fit a new psychographic category that *Wired* named "digital vanguards."[87] Described in media kits as the "consumer paradigm of the 1990s," this audience segment was defined as "venturesome, idealistic, and secure in themselves": "they are the first to adopt new ideas, attitudes and technologies that strike them as right."[88] As the first computer magazine to court consumer advertising, *Wired* made the case that even nontechnological brands would find it beneficial to reach these opinion leaders; lifestyle magazines, it was suggested, delivered undesirable "bottom-feeders" and "technological illiterates."[89] There was a recursive pattern to this positioning. As Paul Keegan observes, "The genius of *Wired* is that it makes the Digital Revolution a self-fulfilling prophecy, both illuminating this new subculture and promoting it—thus creating new demand for digital tools, digital toys, digital attitudes."[90]

Conceived as a magazine that "feels as if it has been mailed back to the present from the future,"[91] *Wired* was presented as a prototype of itself, a preliminary model of the future that it in turn would help create. Forged from the figures and ideals that contributed to the *Whole Earth Catalog* and the WELL, Turner describes *Wired* as another example of a "network forum," a collaborative venue in which dispersed groups could come together and develop local "contact languages" (i.e., "electronic frontiers" or "virtual communities") to facilitate the exchange of ideas and develop shared ways of imagining computing.[92] Media-based network forums, he suggests, are not only partly built from these new languages but also serve as a site for their display. Whereas Turner focuses largely on the rhetorical and conceptual frameworks that network forums offered to writers and journalists, I emphasize the significance of design, which played an equally important role in imagining this digital revolution.

The challenge facing Plunkett and Kuhr was to make visible an "emerging, fluid, non-linear, asynchronous, electronic world" using the "old media" of ink on paper.⁹³ To tackle this, Plunkett turned to a friend at Danbury Printing and Litho in Connecticut, a commercial printer that had just acquired a six-color Harris Heidelberg press. Most magazines were printed on a four-color press, but the Heidelberg press allowed *Wired* to experiment with metallic and fluorescent inks, yielding a brilliant Day-Glo palette that was combined with a visually complex, layered design approach that aggressively signaled the coming of a new technological age. The idea was "envisioning many messages being delivered simultaneously and trying to contrast that with the traditional linear feed of the three-column print formula," says Plunkett. "It was trying to say there's an electronic medium coming, but in the meantime you have to read about it in print."⁹⁴

Inspired by Marshall McLuhan and his collaboration with graphic designer Quentin Fiore, *Wired* incorporated the idea that "the medium is the message" into the look and feel of the magazine.⁹⁵ Each issue opened with a multipage double spread illustrating one quote from that issue. These "mind grenades," as they were called, packed a frenzied and kinetic visual punch.⁹⁶ The premier issue juxtaposed the opening lines of McLuhan's book against a visual collage designed by Erik Adigard (see figure 1.3).

By all accounts, *Wired* was a magazine designed to make people feel like they were missing something important. By covering the "digital revolution," however broadly conceived, *Wired* was able to package different developments—information superhighways, interactive television, VR, the internet—as a cohesive moment, bound by a common "look and feel." Although the magazine came to be synonymous with internet culture, the revolution that Rossetto and Metcalfe first anticipated was not networking but desktop publishing. Through a particular combination of editorial, consumer advertising, and graphic design, *Wired* was able to merge diverse developments into a universal information space, much the way the web and the Mosaic browser subsumed different networks and protocols into a shared window that could help users anticipate the future.

Almost immediately after *Wired*'s debut at the Macworld Expo in January 1993, it became a cultural touchstone that introduced the idea of high-tech geek chic to a mass audience; its neo-Day-Glo aesthetic came to define the cutting edge of digital expression. As mainstream-media coverage from the year reveals, 1993 turned out to be a significant

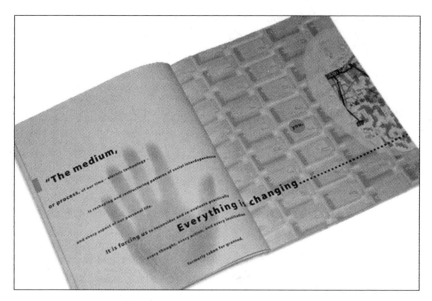

Figure 1.3. Spread from *Wired* 1.1 (January 1993). Design by Erik Adigard.

turning point in the development of a popular techno-cultural imagination. Even those who found its tone too self-congratulatory or its pages unreadable typically described it using the easiest shorthand available: *Wired* was very cool.

Conclusion

Tracing the various developments of the early 1990s that helped forge a new media imagination, this chapter has examined the ways that discourses converged to make the internet, and particularly the web, the premier site in which internet speculation would thrive. Contrary to economic accounts of speculative bubbles, we can see that displacement was much more than a single shock to the macroeconomic system; instead, it was a series of negotiations, bets, and uncertainties that helped bring into relief the contours of a new future.

In particular, 1993 and 1994 were formative to the shaping of a new conjuncture, in which tensions congeal and layers of uncertainty begin to harden into an agreed-on thing, a sense of conviction. The new media future was in fact already here—not as a planned, orderly interactive television system or an immersive VR package but as a crude, alluring mode of computer-mediated communication. During this time, these tensions and uncertainties—between the popular and the elite, difference and universality, fragmentation and consensus—were collapsed and fortified into a working understanding of the digital revolution. Within these cyberspaces, the social imaginary emerged as a kind of "symbolic template" or "cultural conditioning" that generated a sense of identity and inclusiveness.[97] As part of this process, categories of insiders and outsiders were reorganized—or, using the new media jargon of the mid-1990s, there were those who "got it" (*Wired* magazine) and those who did not (old media).[98]

Displacement involved the relocation or reassignment of a number of popular discourses: the "information superhighway" was relocated from the television set to the computer screen; cyberspace was reimagined from the hypergraphical space of VR to the "frontier" territory of computer-mediated communication; the notion of the "digital" was reinscribed from a characteristic of computing technology to a lifestyle, an attitude, and a state of mind. Meanwhile, the web, through the Mosaic browser, moved from an abstract information space to a window through which mainstream users could access and "see" cyberspace. As Arjun Appadurai observes, "the imagination has become an organized field of social practice, a form of work (in both the sense of labor and culturally organized practice), and a form of negotiation between sites of agency (individuals) and globally defined fields of possibility."[99] It is within this organized field of social practice that speculation meets web design. The next chapter, then, takes this up in more detail by examining the ways that the seemingly elusive qualities of "cool sites" helped stitch the emerging new media imagination to the making of the commercial web.

2

Cool Quality and the Commercial Web (1994–1995)

By late 1993, the web consisted of roughly 500 web servers; one year later, this number had ballooned to 10,000, a stunning 1,900 percent surge.[1] As the web grew throughout 1994, navigating the decentralized, nonhierarchical hyperspace of linked resources was a compelling but chaotic experience. The web made clicking on links easy, but one of the biggest frustrations in the days before search engines was the trouble users had to go through to find what was out there and worth visiting. One reporter noted at the time, "There are so many sites, and so many hypertext links between them, that looking for anything is like wandering in some horrendous multi-dimensional maze."[2] Where, in other words, were the shortcuts to the "good stuff"?

As chapter 1 noted, hypertext fundamentally transformed the way users could navigate the abstract spaces of connected networks by making any online resource equally reachable with a single mouse click. This concept of the "equality" of all documents ensured that no link was technically easier to reach or hierarchically valued above other links, but this did not preclude the development of other systems of value, formed precisely through the understanding that some links were indeed more special. Link collections—web pages filled with long lists of hyperlinks to other notable destinations on the web—therefore were seen as valuable content in and of themselves. This was something that the web's inventor, Tim Berners-Lee, realized early on when he started the World Wide Web Virtual Library in 1991.[3] Initially developed as a listing of all known web servers, the links soon became so unwieldy that they were eventually separated into distinct subject catalogs, maintained by experts in these fields who volunteered their services.[4] This collaborative effort, along with its roots at CERN, helped the WWW Virtual Library earn its reputation as one of the web's first valuable and authoritative collections of resources.

For users of the popular Mosaic web browser, there were two other indexes of value that helped users discover new web servers and return to those sites deemed useful, relevant, or interesting. These represent different models of organizing the early web experience: the first, exemplified by Mosaic's own "starting point" for web browsing, the "What's New with NCSA Mosaic" page, was comprehensive and current; the second, which emphasized selection, could be found in the "hotlists" feature—later known as "bookmarks"—built into the Mosaic browser itself. These organizational systems provided an early infrastructure that spoke to two emerging signs of quality: "what's new" and "what's cool."

By making the NCSA Mosaic website the default opening page that loaded each time a user fired up the Mosaic browser, the development team created a well-trafficked hub that offered links to technical documentation and web tutorials. Starting in June 1993, the Mosaic team regularly posted links to new web servers as a way to document "recent changes and additions to the universe of information available to Mosaic and the World Wide Web."[5] Cataloging as many new sites as possible, "What's New" was not organized by subject area but by date; each new announcement appeared at the top of a regularly updated reverse-chronological list. In the beginning, Marc Andreessen and Eric Bina personally assembled these short announcements (e.g., "A Web server is running at the National Radio Astronomy Observatory").[6] But with swelling web activity, guidelines were in place by 1994 for users to submit their own announcements. No priority was given to the links except that the newest appeared at the top: "Any page anyone put up on any topic, we would highlight," Andreessen recalls.[7] By privileging newness over other assessment criteria, "What's New" showed the web "in process," highlighting its growth and incessant activity. As one of the most popular ways of discovering new websites, "What's New" was honored as the "Most Important Service Concept" at the First International Conference of the World Wide Web's "Best of the Web '94 Awards."[8]

While "What's New" chronicled the broader evolution of cyberspace, hotlists introduced a model for organizing the early web experience as a practice of selection and curation. After Mosaic users clicked a "hotlink" to a page they might want to revisit in the future, they could choose to add the link to their personal "hotlist" of saved destinations (see figure 2.1).

This element of customization made web browsing more personalized. But as personal home pages began popping up with more regularity in 1994, users would often publish the output of their hotlists as a service

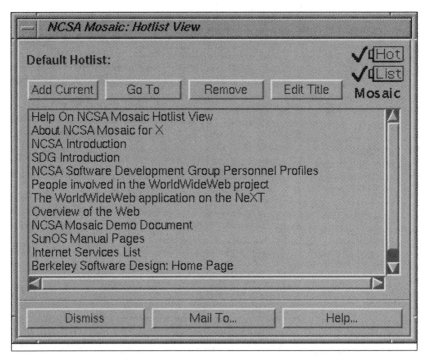

Figure 2.1. NCSA Mosaic's "Hotlist" view. Later, the Netscape browser popularized the term "bookmarks," which suggests a more private collection of links. Hotlists, on the other hand, were commonly shared.

to other users.[9] Because hotlists were frequently shared, they also offered the first collective framework for signaling the "best" or "most useful" pages. Hotlists of other people's hotlists quickly began to accumulate. As the human-factors engineer Jakob Nielsen reflects on this period, "This was the year of home pages that were no more than glorified hotlists with long bulleted lists of links. At this relatively early stage of the Web, people were still easily impressed by anybody who had real, *useful* content."[10] "Useful," as we will see, was one of the most common ways to evaluate a good website, despite (or perhaps due to) its ambiguities.

One such "useful" hotlist started out as "Jerry's Fast Track to Mosaic," a compiled list of favorite links created by a couple of Stanford grad students. By early 1994, the directory was renamed "Jerry's Guide to the World Wide Web," and the curated links, organized as nested subtopics within a hierarchical subject guide, began attracting attention outside the walls of Stanford. By the end of that year, the site—now called Yahoo!—had become a huge success, surpassing 100,000 "page views" a day (up from a few thousand in May); it attracted a million daily visitors just a year later.[11]

In the midst of growth and chaos, the work of collecting, refining, and publishing links to "the best" or "most useful" internet destinations became big business. Phone-book-sized printed web "directories" and "tour guides" hit the market in 1995, offering offline access to "hand-picked" and "noteworthy" links, collected specifically to introduce newbies to the wonders of the web.[12] As it could take several minutes for modem users to load the contents of a web page, these guidebooks were designed to ease the frustration of navigating the "horrendous multi-dimensional maze" by providing a handy desktop reference that pointed directly to "the best of the best" sites.[13] The bulk of the web around this time consisted of content produced by academic, government, amateur/enthusiast, and nonprofit interests: galleries of fractal graphics, weather

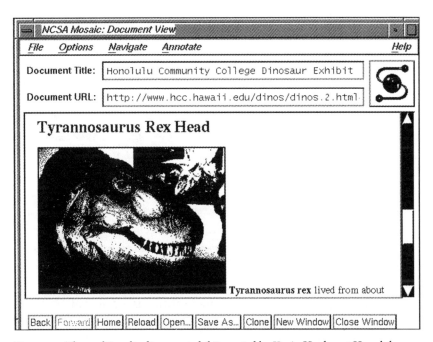

Figure 2.2. The multimedia dinosaur exhibit created by Kevin Hughes at Honolulu Community College was so popular that the site was hard-coded into Mosaic's default hotlist menu.

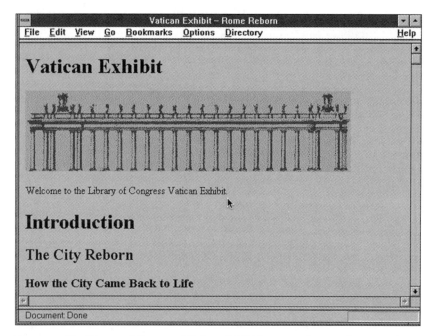

Figure 2.3. The Library of Congress "Vatican Exhibit" presented scanned manuscripts, art, and maps from the Renaissance era.

maps, NASA images of outer space, Kool-Aid fans, Star Wars trivia, the WebLouvre museum, multimedia dinosaur exhibits, hypertext versions of classic literature, the "Vatican Exhibit" (see figure 2.3)—these were just a few of the sites that were regularly circulated through hotlists and favorite-links collections. Museum exhibits were common "favorites."

But the imprecision of these particular benchmarks—best and useful—led these values to be bundled with other types of evaluative criteria that could better express use value not as a characteristic of content but as a condition of the web as a whole. *Web.Guide* (1995), for instance, helps users discover "entertaining, useful, fun, and informative sites on the World Wide Web,"[14] while *Walking the World Wide Web: Your Personal Guide to the Best of the Web* (1995) presents a tour guide of "the most useful, the most beautiful and most advanced WWW sites."[15] Neither text, however, tries to break apart these categories or specify exactly which sites are useful as opposed to fun. Instead, these categories tend to work in the service of one another, painting a picture of the web as a vast and diverse multimedia publishing space with exhaustive breadth and depth, an information repository with loads of "free stuff" available for the taking, and a shared, collaborative space for connecting, communicating, and contributing. These intersecting values, broadly understood as "fun," "useful," and "participatory," comprised a discourse of "cool links" to "quality" internet destinations that could help newbies grasp the power of the web.[16]

The Quality of Cool

So far, internet studies has tended to be less reflective about the politics of "quality" than television studies, which has spent decades interrogating the power relations that arise in discussions of quality television. Charlotte Brunsdon observes in her seminal 1990 article "Problems with Quality" that "any interrogation of what is, and could be, meant by 'quality' in a discussion of television" involves "discourses of judgment."[17] Too often, Brunsdon contends, "quality" is eschewed by media and cultural studies, where the impulses that drive the study of popular culture are also those that repudiate the class and cultural privileges that have defined the history of high art. Drawing on Bourdieu's critique of taste in *Distinction*, she argues that there are always issues of power at stake in notions such as quality and judgment: "Quality for whom? Judgment by whom? On whose behalf?" Invoking subjective factors as the ground on which judgments of quality cannot be generalized, she argues, "blurs the role of structural and institutional factors in the formations of these judgments."[18]

Throughout 1994 and early 1995, various sets of evaluative criteria—"discourses of judgment"—were deployed in the service of distinguishing "quality" web pages: authoritative and expertly vetted (WWW Virtual Library), comprehensive and logically organized (Yahoo!), actively updated with new content ("What's New with NCSA Mosaic"), and shared collections of personal favorites (hotlists). At the same time, discourses of quality also circulated around the work of making the web. HTML tutorials and other early how-to books on web publishing, for example, stressed structural virtuosity above all else. According to Laura Lemay's bestselling 1995 book, *Teach Yourself Web Publishing with HTML in a Week*, a quality "web presentation" is a well-crafted one, with

Figure 2.4. Examples of "cool" award icons (1994–96).

sound code, logical organization, context-specific hypertext links, and the appropriate use of HTML formatting (headlines, bolding, horizontal rules, lists) to aid "scannability." Clear writing, descriptive titles, and appropriate navigational menus are prioritized over graphical sophistication. Lemay urges readers to reduce the number of images and other visual elements in their pages and explicitly states, "You shouldn't take design into consideration when you write your web pages."[19] This structural approach to quality web authoring emphasizes standards compliance and cross-browser functionality. As we will see in chapter 3, various groups sought to deliberately downgrade these values in favor of established graphical design principles once the dot-com boom was in full swing; only after the bubble burst did these structural markers of quality once again surface as a dominant design discourse (see chapter 5).

However, the sites that received the most favorable reviews in web tour guides, the popular press, and magazines such as *Rolling Stone* were rarely hailed for their structural virtuosity or their "scannability." Rather, these sites were customarily celebrated through the more elusive cultural shorthand of "cool." From media descriptions of the internet in

1994 and 1995 to the vernacular rhetoric of hotlists and personal home pages, the early web was teeming with declarations of "cool"—Cool Links, Cool Picks, Way Cool Websites, Cool Site of the Day, Cool Surf Spots—that stuffed the virtual landscape like a trophy collection that had outgrown its case. Recognition by well-respected cool-links lists was a welcomed honor, one that was visually marked on cool sites with graphical status icons that bestowed a stamp of "quality" (see figure 2.4). Esteemed sites that collected enough of these accolades would often erect new "awards.html" pages to house the entire collection.

But what could "cool" even mean amid such ubiquity? Thrown around so much that it has come to be an all-purpose descriptor for anything even loosely agreeable, "cool" suggests both value and ambiguity. As such, it may seem a tenuous place to start an inquiry into the web's industrial logics, aesthetic sensibilities, and modes of production. But, I argue, the early web's preoccupation with cool might be read in ways other than as a vague category of dubious distinction. As Alan Liu points out, cool is a historical condition, one that can be elucidated through cultural criticism and analysis.[20] Before dismissing "cool sites," we might instead search out cool as the register of a new structure of feeling expressed via the web's nascent visuality: its design, tone, and modes of address.

The Work Cool Did

By far the most detailed treatment of the phenomenon of "cool" in relation to information culture is Alan Liu's *The Laws of Cool: Knowledge Work in the Information Age* (2004). Vast, challenging, and ambitious, the book weaves a critical history of knowledge work and corporate culture in the twentieth century, from Taylorist efficiency principles to mainframe computer culture, from white-collar data-entry jobs to the "user-friendly" interfaces of information technology (IT) work. In this account, postindustrial cool represents the cold face of alienation within 1990s cubicle culture, which is "both strangely resistant to and enthralled by the dominating information of postindustrial life."[21] Within the logic of cool sites, Liu finds a self-conscious awareness of information, technology, and the interface. At the same time, cool represents a paradoxical resistance to information by reveling in a certain uselessness: high tech just for the sake of high tech, mundane chronicles presented to a global audience, random link generators that convert relevance into play.[22] Expressed as "information designed to resist information," cool emerges in *Laws of Cool* as an outlet or coping mechanism for those who live and work inside the system of postindustrial capitalism. From within this regime of knowledge work, the cubicle subject seems to say, "We work here, but we're cool. Out of all the technologies and techniques that rule our days and nights, we create a style of work that is 'us' in this place that knows us only as part of our team. Forget this cubicle; just look at this cool Web page."[23] Liu uses the refrain "We work here, but we're cool" to encapsulate the culture of information and the ethos of modern cubicle alienation through an affect that both humanized and disengaged from the system of technological rationality. But for all the detailed attention to the character of "cool links" and "cool sites of the day," postindustrial cool comes across as a single enduring phenomenon that barely distinguishes between web content produced between 1994 and 2000. Surely the cool of 1994 was not the same as the cool of 1997, which was not the same as the cool of 2000? As we will see in later chapters, the "rules" of cool were constantly rearticulated to shifting definitions of quality as the commercial web and the cultural industries that assembled to produce it grew by gargantuan portions over this six-year period.

To unpack these distinctions, we might instead consider *the work cool did* for some of those first measured attempts to bring commercial culture to cyberspace. While Liu locates cool within the broader institutional pressures of knowledge work, I contend that there is also much to be gained by attending to the "small chronologies" of cool, the intricacies that take place on the ground floor as "cool"—an aesthetic, an evaluative criteria, and a corresponding set of production practices—changes in response to rapid developments within an emerging commercial web industry that formed alongside the internet speculative bubble.[24]

In chapter 1, I described displacement as the first stage of a speculative bubble, formed through the conjuncture of distinct currents that enabled a different vision of the interactive future to be imagined. "Cool," I suggest here, is the web's modality of that displacement. "Cool" gestures toward shared meanings about the nascent web, how its "look and feel" was socially experienced and collectively expressed. In this way, the discourse of cool, and its materialization in the web interface, helped construct a social imaginary that cohered around loosely agreed-on (though not easily articulated) principles, recommendations, and hierarchies of value. These shared assumptions, I should point out, lent crucial support for a more widespread recognition of the web's economic potential and hence for the speculative transition from "displacement" to "boom" by the fall of 1995.

In the next section, I examine cool's articulation to quality during this moment of displacement, first as the register of a new visuality and structure of feeling and then as a set of production practices and aesthetic sensibilities that provided an emerging web industry with the strategic devices for managing the commercial incursion of cyberspace. Turning to three case studies of early commercial websites launched between the fall of 1994 and the summer of 1995, I conclude by examining how "cool quality" was operationalized through strategies that embraced storytelling, participatory culture, community building, and gift economies. For those marketers that were among the first to stake a claim in cyberspace, "cool quality" helped commercial interests "fit" within the rhythm, attitudes, and values that defined the look and feel of the early web.

The Look and Feel of Cool

> The magic of a cool site is difficult to define. Much of it is that sense that the whole is the greater than the sum of its parts. It's the way the content suddenly seems "made for the Web." It's that indefinable *it* that lifts one site above the ordinary.
> —Teresa Martin and Glenn Davis, *The Project Cool Guide to HTML*

According to *The Project Cool Guide to HTML* (1997), cool sites are cool because they suddenly seem "made for the Web." As such, they can reinforce or unsettle shared experiences of the network, perceptions of value, and collective visions of the future. Cool sites speak the web above all else and are thus key objects with which to think about its design, experience, and meaning. If we are to try to make sense of cool as a way of talking and thinking about the web, as a set of evaluative criteria for identifying "quality," we need to begin by considering how cool is articulated to the web's look and feel. As the mode in which displacement is expressed, cool conveys a collective tone, ethic, and pulse of the web; it articulates the status of a single site to the "look and feel" of the web as a whole.

This phrase, "look and feel," first gained traction in intellectual-property lawsuits in the 1980s and 1990s, in which the visual appearance of a product or its packaging (known as "trade dress," the composite of elements taken as a whole that consumers recognize as being made by a particular company) could be protected under trademark law: the distinctive grille of a Rolls Royce, the familiar green glass bottle of Perrier, or, as US courts ruled in 2012 in *Apple v. Samsung*, the iconic Apple iPhone. In the late 1980s, Apple famously challenged Microsoft and Hewlett-Packard for stealing elements of its visual display in what became known as the "Look and Feel lawsuit," but the case was thrown out in 1993 due to the complex issues around licensing agreements.[25] Today the phrase is commonly used to talk about the design of interfaces (hardware, software, graphical layout, color, shape, behavior, navigation, interactive style, etc.), from mobile devices to television menus, remotes, personal computers, the software that runs on those computers, and of course, the design of websites.

Here, though, I approach the "look and feel" of the web in a far more expansive sense. Although I briefly address each individually, the point of "look and feel"—in the ways that both legal teams defend it and software and web designers understand it—is that the two terms depend on each other to constitute a formative whole experience and cannot therefore be easily reduced to a series of elements. With that said, I take the "look" to be bound up with visuality and its links to speculation and the imagination. In the introduction, I discussed the historical conditions of these connections as visuality became hinged to the history of capital. Visuality, remember, is not the same thing as vision. The field of visual culture, as it emerged in the 1980s (most notably with Hal Foster's 1988 edited collection *Vision and Visuality*) and expanded through the 1990s, made a distinction between the physiological process of seeing and the "social fact" of visuality, which plays a key role in the production of subjectivity. According to Nicholas Mirzoeff, "visuality is very much to do with picturing and nothing to do with vision, if by vision we understand how an individual person registers visual sensory impressions."[26] Not simply composed of visual perception in the physical sense, visuality is instead a process "formed by a set of relations combining information, imagination and insight into a rendition of physical and psychic space."[27] In this way, the web's visuality goes beyond the mere visual design of the interface to include how an abstract information space is produced, navigated, and imagined. It was the visuality of the web—not simply its graphical representation—that helped people "get it," to understand that something fundamentally different was in the air.

Intimately bound up with the web's visuality is the "feel" of the web, which I approach by way of Raymond Williams's notion of "structures of feeling." In developing this concept, Williams was looking for a way to talk about a particular quality of lived experience in an emerging present, "a kind of feeling and thinking which is indeed social and material, but each in an embryonic phase before it can become fully articulate and defined exchange."[28] As Williams describes it, a structure of feeling is a social formation that is "structured," but as a social configuration that has not yet fully formed, it exists "at the very edge of semantic availability." So he looks to art and literature for glimpses of our always-emerging world; it is here that we can find the "specific rhythms" or the "characteristic elements of impulse, restraint, and tone" that characterize the ongoing process of social change. He writes, "The idea of structure of feeling can be related to the evidence in art and literature—semantic figures—which are often among the very first indications that a new structure is forming."[29]

Whereas Williams turns to texts such as the industrial novel of the 1840s to show how a particular structure of feeling emerged from industrial capitalism, I look to the "anthologies of cool"[30] that circulated around early (1994–95) "Cool Site of the Day" links, "Cool Links" collections, and books such as *1001 Really Cool Web Sites* (1995), *Creating Cool Web Pages with HTML* (1995), and *CyberHound's Internet Guide to the Coolest Stuff Out There* (1995).[31] What such an exercise helps us see, I argue, is an emerging new media imagination as it forms alongside the commercial imperatives of capital. Cool surfaces in these accounts according to clear patterns that are generally agreed on, yet, as the epigraph that opens this section states, "the magic of a cool site is difficult to define." It is "that indefinable *it*" at the edge of articulation that struggles to express exactly what this inchoate "it" is.

Cool Links

An emergent discourse of "cool" was already present in early link collections and published hotlists, but it was given a very visible boost in

August 1994 when Glenn Davis, a project manager at the Norfolk, Virginia, internet service provider (ISP) InfiNet, began sharing links to cool sites he found online to a "Cool Site of the Day" (CSotD) page (see figure 2.5) on the company's website.³² The site made Davis one of the first internet celebrities, and CSotD (pronounced *SEE-sought-DEE*) became known as the web's "arbiter of taste."³³ The site was announced on a number of UseNet newsgroups, including comp.infosystems.www.misc (a forum for general discussion of the web) with the following: "Need a daily web fix of something new? Try The Cool Site of the Day. Every night at midnite the Cool Site of the Day gets set to point at a new Cool Site. You'll never know what's there until you take the link so expect to be surprised."³⁴ With this, "cool" was serialized as a daily hyperlink to somewhere new, a ritual for satisfying one's requisite "web fix." At the heart of CSotD was the "surprise" element of the cool link. Browsing the archive of 1994–95 Cool Sites of the Day, one might surmise that the element of surprise is partly related to the vast range of content featured—a hodgepodge so arbitrary that it appears hard to draw out any defining characteristics of cool.³⁵ In March 1995, for example, the links feature sites as diverse as H's Home Page, the Thai Heritage Project, Coors Brewing Company's Zima website, Cybergrrl, and the UC Museum of Paleontology. Cool sites are clearly not of a single genre, nor do they implement technology or conceive of content in exactly the same way. They are commercial and noncommercial, produced by amateurs and "professionally" designed, and they include both single web pages and large sites sprawling with content. Faced with such incongruities, one might wonder if there can even be a single logic of cool. Yet, as Alan Liu notes, browsing early cool-links lists, one finds "a surprising amount of consensus. . . . Collectively they attest to consistency in the kinds of sites deemed cool."³⁶

Taken as a collection, cool sites were those in tune with the ethos of the web; like the web itself, cool-links collections shunned traditional hierarchies by embracing the "equality of all documents" and insisting that any site, not just well-funded or institutionally connected ones, could potentially be cool: movie trivia sites, fish-tank webcams, mortgage calculators, email syntax guides, frog-dissection simulators, NASA image galleries, clip-art collections, the White House "interactive citizens' guide," Tori Amos fan pages—all were cool. Yet this inclusive view of cool subsequently introduced a new hierarchy of value: while cool

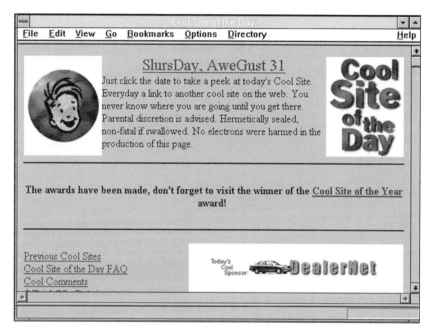

Figure 2.5. Cool Site of the Day (1995).

was an honor within reach of potentially any website, only those sites that spoke the web as a whole, that exemplified its spirit or helped users to imagine the scope of its potential, were explicitly identified as cool. In other words, it was precisely through these arbitrary juxtapositions that the contours of cool discourse could be understood as articulating a new structure of feeling. Cool sites gestured toward the sheer range of things the web could be: its temporal and spatial dislocations, its distinction from prepackaged mainstream media, its promise as a multimedia vehicle for self-publishing, the incredible blend of personal, mundane, and extraordinary encountered in the course of surfing the web.

In the web's first few years, before the coherence of an organized new media cultural industry and, indeed, before the social experience of the web was fully processed or habituated, the new temporal and spatial experiences introduced through Mosaic's window to the web were vividly described by early users as "thrilling, disorienting, unreal."[37] Consider the following electronically published account, titled "Justin in the Zone," by the journalist Gary Wolf, who was working with Justin Hall (a

Swarthmore undergrad who gained early notoriety through his personal site, Justin's Links to the Underground) and the launch team of *Wired* magazine's new online publication, *HotWired*, in September 1994 (it was distributed to *Wired* staff via email and published online by Howard Rheingold):

> With Justin at the mouse, we distributed our brains between the five or six Mosaic windows that were always open, clicking on new links while waiting for others to load and startled by an endless sequence of impossible juxtapositions. Justin urged us to follow a link labeled "torture" while an animated diver performed background somersaults and an anonymous Indonesian artist gave us a glimpse of demonic possession. The images went by so fast, and yet so disconnectedly, that time itself seemed to be jerking backwards, as if Mosaic were a sort of temporal strobe-light creating the illusion that the evening was flashing before our eyes in reverse.[38]

With Wolf's and Hall's brains "distributed" across a number of Mosaic windows, Wolf's account offers a sense of an emerging collective consciousness that is startled by these "impossible juxtapositions" encountered as they move through content that is dark, silly, foreign, familiar, and historical. There is a new temporality (the beginning of internet time?) to this experience; as the group moves onward to the next link that combs backward through old audio archives, a multidirectional mobility produces the sensation that the evening was "flashing before [their] eyes in reverse." Tara McPherson describes the sensation of freedom, choice, immediacy, and movement associated with web browsing as "volitional mobility," which emerges from the way that the web structures a sense of causality in the act of navigation, "a feeling that our own desire drives the movement."[39] That is, as we are propelled along by the search for what is next, our choices seem to structure the web's spatial and temporal dimensions, "permitting the Web surfer to move back and forth through history and geography, allowing for the possibility (both real and imagined) of accident and juxtaposition" that we feel we create for ourselves.[40]

"Cool" sites, then, often found ways to exploit the "feeling" of the web by putting new spatial and temporal experiences on display. The first webcam, connected to the web in 1993, famously pointed to a coffeepot at the University of Cambridge so that computer scientists in the building could check to see if the pot was full before making a trip down the hall or up a flight of stairs. Although the slowly loading series of grainy black-and-white images of a coffeepot was technically "useless" to the majority of internet users not located in Cambridge, the site became an early landmark that demonstrated the power of the telepresent gaze, a novel experience of visuality from afar. The fact that most visitors to the site were located too far away to use the site as intended only made the site that much cooler; it was a destination that threw the whole web into sharp relief by instantly making the power of the web visible.

The CSotD team tried repeatedly to give advice on how to create a cool site, but the explanations were so imprecise that, finally, the chemical compound "coolium" was invented on the site's frequently asked questions (FAQ) page to explain the uncertainty of it all: "For years, scientists have sought the answer to life's most fundamental question: What is cool? . . . Can I create cool? What are the ingredients of cool? . . . We did discover that all coolness shares one thing in common: an actual chemical element that we call Coolium." To create "coolium" and be selected Cool Site of the Day, the editors offer the following advice: First, "create something useful," but "if you can't be useful, be entertaining." Next, "be aware of bandwidth." And if all else fails, they suggest, "surprise us."[41] Hence the "magic" of the cool site, which, as the CSotD editor Richard Grimes (who succeeded Davis in November 1995) admitted, comes down to a "gut feeling" about the experience of the web and how it works. "If I go to the site and I am hit in the stomach with the need to explore the whole site, if the site takes my breath away, then it's cool," he explains.[42] Consulting other "cool" guides such as *1001 Really Cool Websites*, *CyberHound's Internet Guide*, abandoned hotlists online since 1994,[43] and numerous cool-links collections, we encounter a series of intersecting qualities of cool—variously explained as fun, useful, interesting, entertaining, weird, informative, visually appealing, or technically advanced—that gesture to a set of shared meanings and expectations about the web.

Take "useful," one of the most common attributes of cool. As the user (and the work of web design) was constantly reimagined throughout the course of the web's first decade, corresponding design discourses of "usability," "user-centered," "user-friendly," and "user experience" emerged to translate and encode these values in the design of the web. Likewise, early discourses of cool sites as "useful" represent, above all, a concern for the user; the admonition to "be aware of bandwidth," for example, acknowledges the amount of time that modem users spend waiting for a site to download. The Coolium FAQ advises that, to create something useful, web publishers should "make the wait worthwhile by providing some kind of service." "Service" is understood in a number of different ways. Recall that the "What's New with NCSA Mosaic" page was named "Most Important Service Concept" in the spring 1994 "Best of the Web Awards." Here, the service helps people find new resources on the web. Navigation systems, hotlists, annotated link collections, search engines, and random link generators all provide similar services.

But simply providing information was also understood as a useful service, as long as the cool site takes advantage of the web's encyclopedic capacity for depth and detail. In *1001 Really Cool Websites*, "informational coolness" is explicitly defined by "depth of data," which is identified as a key criterion for identifying "really cool" sites for inclusion. "The web enables the presentation of vast amounts of data, and thus makes sites that offer minimal amounts of information on topics they purport to 'cover' seem all the more trivial and absurd," explains the author, Edward Renehan.[44] The immense multimedia materials of a virtual museum exhibit or a collection of detailed tutorials on web production with HTML, for example, are cool because they use the web to provide greater scope than would be possible through traditional media channels. There are many other types of useful web and internet services that appear with frequency in cool-site collections as well: searchable databases (of movie titles, Supreme Court rulings, Bartlett's familiar quotations), computer-mediated communication services (forums, chat, multiplayer MOOs and MUDs), live information (weather services, stock data, world time zones), computational tools (financial calculators, mathematical simulations, fractal generators), or "free downloads" (clip-art collections, software, games, sound clips). "Useful" was indebted to the internet's gift economy—the free sharing of information and expertise—for social rather than economic capital.

While useful sites were cool because they provided users with the ability to do something with the web, it was the capacity for "entertainment," for fun, that made the web popular. Limited by browser technologies, network speed, and early HTML standards, the websites of 1994 (as viewed in Mosaic) displayed blue hypertext links on a gray background with content arranged top to bottom on the screen with little more than headlines, bullet points, and horizontal rules to divide the information. It is worth noting, however, that for new web users, even the "text-only" web appeared strikingly different from earlier internet experiences. As *Web.Guide* explains, "It makes a huge difference just to have titles that appear in larger type than the rest of the text. It's nice to see italics and boldface for a change. And the inclusion of even a small picture can change the whole feel of a page of information. . . . A circus has been loosed in a world of damp, gray typescript."[45] In a primarily text-based environment, writing carries significant weight; "fun" was expressed not just in the pleasure of gift economies but through a site's "personality," its voice, tone, attitude, and direct, informal mode of address. Sites that deployed clever, witty, sarcastic, or goofy language did not take themselves too seriously; they made it clear that there was another person out there, on the other side of the web page. Personality, therefore, revels in the quirky, intimate, ironic, casual, and open attitude of internet culture. It surfaced in copy (CSotD's "Slursday"), visual design (clip art, psychedelic graphics, animated gifs), and in the fascination with "personal home pages" and their characteristic blending of professional and personal, public and private identities.

By providing personal glimpses of virtual strangers, home pages juxtaposed the private domestic sphere with the official address of the public profile. In an October 1994 issue of *Wired* magazine ("Why I Dig Mosaic"), Gary Wolf reflects further on the particular pleasures and distractions of using Mosaic to browse the web. Having wandered off track from his intended mission to the CERN site, Wolf aimlessly pursues a poetry archive when he stumbles onto the home page of a physicist interested in string theory. Alongside obligatory links to research documents, Wolf notices a link to "Benjamin's Home Page" and discovers not a scientific co-author but a "beaming, gap-toothed 3-year-old, who announced at the top of the page that his research interests were, 'Sand. Also music, boats, playing outside.'"[46] At this point, Wolf explains, "I began to experience the vertigo of Net travel." He describes a vivid sense of dislocation that is linked to the web's peculiar form of visuality: "It was a type of voyeurism, yes, but it was less like peeking into a person's window and more like dropping in on a small seminar with a cloak of invisibility. One thing it was not like: it was not like being in a library. The whole experience gave an intense illusion, not of information, but of *personality*. I had been treating the ether as a kind of data repository, and I suddenly found myself in the confines of a scientist's study, complete with family pictures."[47] Highlighting the "personal" in personality, Wolf's web account emphasizes the role of people, the social dimension of these new spatial and temporal experiences. Williams's concept of "structures of feeling" offers one way to think about these affective experiences as they were lived, "in process" and not yet fully formed. Although not yet fixed, institutionalized, or molded into recognizable forms, such feelings, Williams argues, are not merely personal, idiosyncratic, private, or subjective; they are still, importantly, "social experiences."[48] As such, they make possible the generation of a social imaginary, one that seems to grope for the language to describe the web's spatial and temporal mobilities before they have been reworked into the commercial web's structured information architecture, intentional navigational schemas deployed as the economy of the link became evident.

Through the shared practice of identifying and collecting examples of "cool," the ambiguity surrounding this semantic expression began to settle on those characteristics that somehow managed to capture the thrill, wonder, and tone of a fledgling medium as it was being constructed and collectively evaluated. Nebulously aligned with such qualities as "useful" and "fun," the cool web was first and foremost a "participatory" experience, collectively defined, circulated, and cross-referenced to arrive at an early, shared sensibility of cool. Describing how cool links were chosen for *CyberHound's Internet Guide to the Coolest Stuff Out There*, the editors admit that "the sites were chosen after scanning various 'cool site' lists, eyeballing the Web at length, consulting professionally with 'people in the know,' and receiving strong vibes from certain URLs."[49] This informal consultation of other people's definitions of cool had been routine since the emergence of hotlists of hotlists. Notably, the very first Cool Site of the Day, announced August 4, 1994, was the Infobot Hotlist Database, a site that collected the contents of various hotlists and periodically produced a list of the most popular documents. Inaugurating the Cool Site of the Day with a link to the hotlist database, Davis's collection (and the cool-links lists that followed) were as much about validating a collective cool sensibility as they were about selecting particular instances of cool's expression in the design of an interface.

As a collective sensibility, the codes of cool played an important role in the social construction of a "quality" web presence. But for the emerging new players that specialized in the design and production of the commercial web, cool also offered a set of strategic devices for easing concerns about a commercial incursion of cyberspace. The next section analyzes how an ethos of "cool quality" figured in some of the first commercial websites as a set of aesthetics and production practices that embraced the characteristics of "useful," "fun," and "participatory." By constructing an online presence within the codes of "cool," marketers and agencies were able to structure their address and architecture within established Net conventions and thus minimize potential backlash to the web's commercialization.

The Rise of Web Industries

Few advertising agencies had much experience with interactive media in 1993, when the interactive television tests (discussed in chapter 1) were preparing to inaugurate the "information superhighway." Although the incorporation of advertising was a major component for interactive initiatives such as Time Warner's Full Service Network, there was a persistent sense that agencies were unprepared for the interactive future.[50] For one thing, interactivity presented advertising agencies with entirely new creative demands: advertising typically flowed one way—from the sponsor to the audience—but now audiences were expected to be in command of information flow, which meant creatives would have to give up their sense of tight control over the advertising message.[51] Not only did this interactive require new skills, but it promised to transform the traditional ways of measuring audiences by estimated audience share. If clients had real-time data showing them exactly who viewed their ads and for how long, agencies could be held to stricter standards. Plus, interactive was expensive. Reeling from the toll that the recession took on the industry, there were looming doubts about precisely how the rules were changing. The last big interactive revolution, the arrival of the commercial two-way communications system known as Videotex, was once "poised to replace cable television as the hot new media toy of

1984."[52] But it never caught on. Many people in the advertising industry could not help but wonder: might the information superhighway be the sequel to the Videotex revolution?

In the spring of 1994, the stakes were raised for advertisers' involvement in new media when three key events helped transform attitudes and strategies for bringing marketers online. Perhaps most significantly, at the annual meeting of the American Association of Advertising Agencies (the 4As) in May 1994, Ed Artzt, chairman of Procter & Gamble, the world's largest advertiser, issued a provocative wake-up call to advertising agencies, urging them to embrace new media technologies or risk becoming irrelevant. This was the first time a giant marketer had addressed the issues raised by the development of the information superhighway. Agencies, Artzt exhorted, must confront a new media future that would not be driven by traditional advertising: "Most of the hype about the 'information highway' has focused on technology, but we're interested—as I know you are—in consumers: What is it that they want from these new services, and how much are they willing to pay for them? . . . The question is, if they get what they want—and I have no doubt that they will—how will that change the way they use television? And what will it mean for advertising?"[53] Reminding agencies how brilliantly the industry once adapted to new technologies such as radio and television, Artzt urged his audience to "grab all this new technology in our teeth once again and turn it into a bonanza for advertising."[54] The effect of this speech on large advertising agencies "couldn't have been more persuasive than if Artzt had held a double-barreled shotgun on the ad execs," reported *Communication Arts*.[55] Former vice presidents of marketing suddenly found themselves with new job titles as VPs of new media, and agencies scrambled to staff up in-house interactive teams, buy up small production houses, form interactive task forces, and write up white papers on "the future of advertising." Although Artzt was talking about a future of advertising that was linked with interactive television, elsewhere another wake-up call jarred the ways that advertisers thought about interactive media.

A month before Artzt's address, the first widespread assault of commercial spam had hit internet newsgroups. Even though legislation had been passed by the US Congress in 1992 opening the internet to commercial traffic in the move toward privatization, the internet community adhered to a strong system of etiquette ("netiquette") and social norms that made the practice of unsolicited advertising a risky endeavor subject to extreme outrage from internet denizens. It was one thing to offer a service in which computer-savvy users could browse electronic catalogs, but blatant advertising was frowned on. So when two immigration lawyers from Arizona, Laurence Canter and Martha Siegel, flooded hundreds of UseNet newsgroups with advertisements for their firm's legal services representing immigrants in the "green card lottery," angry UseNet recipients counterattacked by deluging the offenders with thousands of insulting emails (a tactic known as "flaming"), overwhelming the lawyers' service provider.[56] In an account of the "virtual lynch mob" that formed in response, Peter Lewis writes in the *New York Times*, "Angry mobs pursue Laurence Canter and Martha Siegel from network to network. Electronic mail bombs and fax attacks rain down upon them. Their digitized photos are posted on computer bulletin boards around the world, like 'Wanted' posters on the electronic frontier."[57] Unscathed, Canter and Siegel announced the publication of their book *How to Make a Fortune on the Information Superhighway* later that year.[58] Given the precedent, it was no wonder advertisers expressed trepidation following this well-publicized computer war over commercial spam. Might interactive advertising do more harm than good?

But the impending launch of online magazines such as Time Warner's *PathFinder* and *HotWired* (a web companion to *Wired* magazine) introduced another model for interactive sponsorship. Just as Artzt was igniting the industry with his wake-up call to interactivity, *HotWired* was seeking advertisers to sponsor each section of its website. The biggest advantage that these online publications had over interactive television tests was that advertisers and agencies understood the model of magazine sponsorship. "I've got a real problem today with interactive television technology tests," the director of Ketchum Interactive complained in August 1994. "I can endorse CD-ROM. I can endorse [commercial] online [services]. I can endorse place-based systems and two-way kiosks," he said. "We've done it. It works. But I believe that advertising agencies are doing a disservice to their clients by recommending that they pay to support interactive TV trials. I can't endorse the unknown."[59] Unlike the "unknown" model of interactive TV trials, an ad-supported publishing model on the web was familiar and made sense. Involvement could be justified to clients, and results could be analyzed and measured.

Bob Schmetterer, president of the ad agency Messner Vetere Berger McNamee Schmetterer (MVBMS), had been in the audience during Artzt's speech and decided to jump-start his agency's involvement in cyberspace by making a media commitment to buy ad space for the *HotWired* launch.[60] As the account executive Frank D'Angelo explained, "This initial assignment was under the guise of 'let's explore this new medium and see what happens.'"[61] Four of MVBMS's clients (MCI, Volvo, Club Med, and 1-800-COLLECT) were chosen to be part of that first campaign, although the decision to test the waters of interactive media was so experimental that the ad buy was placed before clients even approved. The major impetus, according to D'Angelo, was that the agency (and of course the agency employees) would gain valuable experience in new media: "Corporate America was still largely unfamiliar with the graphical web, so we didn't even try to sell the concept. We decided to commit agency media and development dollars to place client banner ads on *HotWired* without clients' prior consent or knowledge. The way [Schmetterer] saw it was if they liked it, they would be happy to pay us and if not, that was OK too; but at least the agency would get a running start at exploring this new exciting medium that was on course to change all of our (professional) lives."[62]

HotWired introduced a new model for commercial sponsorship online: the "banner ad." This was a rectangular, "clickable" graphical ad unit that was positioned at the top of a web page and bore a sponsor's message. While the practice of blocking space within media content for commercial sponsorship was well established, the work of developing these banners raised new questions: What would happen when users clicked on an ad? Where would they go? Would sponsors have their own "home pages" on the web, and if so, what would these pages look like? What would visitors actually do there?[63] These questions did not have immediate answers; in fact, the questions themselves only began to emerge in the process of buying ad space on *HotWired*. D'Angelo, who was involved in the media buy for the *HotWired* launch, later recalled, "We were given the ad specs by *HotWired* and it was only then that we realized banners ads were clicked on and could drive consumers to a client designation on the web. Oops! This accidental lesson sparked us to develop websites for these initial ad banner placements."[64]

Meanwhile, small startups that specialized in designing, programming, managing, and hosting websites were quietly making inroads into the agency-client sector. Not to be confused with ISPs that provided network access, this new type of interactive shop was known initially as "Web service providers." As the first fourteen *HotWired* sponsors were secured, the work of designing several of these commercial "mini-sites" went to a still-struggling year-old startup, Organic Online. With web production experience so scarce, Organic found itself one of the few providers capable of delivering potential *HotWired* advertisers on the promise of an effective online presence.[65] The tight relationship between Organic and *HotWired* was a symbiotic one, built from personal connections that linked high-school friends from Wisconsin, computer-networking and "geek" subculture, the local electronic-music and rave scene, and an abiding interest in multimedia, digital art, and technoculture that cohered in San Francisco's South of Market area (SOMA) in the early 1990s. Organic co-founder Jonathan Nelson and *HotWired*'s "Online Tsar," Jonathan Steuer, grew up together in Milwaukee and were old school friends. Two other Organic co-founders, Brian Behlendorf and Jonathan Nelson's brother, Matthew, worked for *Wired* as well as Organic and helped to launch the *HotWired* project—Behlendorf as webmaster and Matthew Nelson as advertising account manager. They split their time between the two companies, a job made easier because Organic, Wired Ventures, and *HotWired* were all located in the same Third Street building in the SOMA area that came to be known as "Multimedia Gulch."

Although the two organizations rented space in the same building and shared talent, there was no formal financial or client arrangement between them.[66] Jonathan Nelson tutored the *HotWired* ad-sales staff on web basics, and in return, the ad staff passed along new customer referrals.[67] Nelson participated in a number of *HotWired* sales calls with the ad-sales staff, and the two groups presented to media buyers together, with Organic pitching the microsites and the *HotWired* sales team talking about the audience.[68] As a result, MVBMS hired Organic to design the commercial home pages for Club Med, MCI's 1-800-Collect, and Volvo.[69]

These early partnerships between web service providers and ad agencies were hardly a marriage of equals. The first wave of companies, such

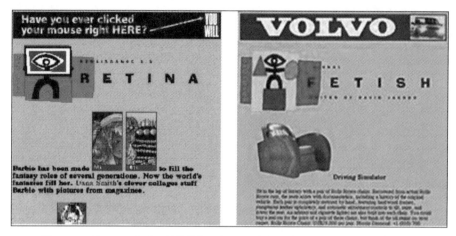

Figure 2.6. Banner ads for AT&T (Modem Media) and Volvo (Organic Online) on the HotWired site.

as Organic (San Francisco), Free Range Media (Seattle), CyberSight (Portland), OnRamp (New York), and K2Design (New York), largely worked as vendors for ad agencies such as MVBMS, Ogilvy, Chiat/Day, and J. Walter Thompson, formatting content and cranking out banner ads and websites according to the agencies' creative direction. "We get small companies pitching us left and right, and we use them for formatting computer language," explained a partner at MVBMS.[70] Since ad agencies were the ones hiring these "interactive suppliers" for their accounts, they had an upper hand in the early stages of the web media industry. There was the vivid sense that professional web design might be a short-lived career. "Pretty soon, we'll be doing it all in-house and there won't be a need for them," noted the MVBMS partner.[71] With the client's ball clearly in the ad agencies' court, web service providers worked hard to cultivate synergistic relationships, repeatedly expressing outright admiration in the trade press for the consumer insights that traditional agencies could deliver.[72] As stewards of the brand, agencies saw these interactive startups as filling the same role as the TV production companies that made their clients' commercials; agencies would maintain client relationships but farm out much of the responsibility to third-party companies.[73]

Two Approaches to Building the Commercial Web

With the internet community's vitriolic reaction to the green-card-lottery spam still fresh on their minds, many advertisers were wary of offending the very community they were trying reach. As a result, some early commercial sites were designed to tread very carefully. The interactive consultant helping AT&T design its consumer-products website for *HotWired* explained the rationale for the site: "We can't call it advertising because of etiquette on the internet that guards against blatant advertising unless you're in a strictly commercial area," such as Prodigy, America OnLine, or CompuServe. "We have to play by the rules and be informative."[74] Shunning direct advertising by instead "providing layers of information," sponsors were mindful of the repercussions that might await those companies caught violating established internet social conventions.

Such anxieties are far from new; indeed, similar apprehensions accompanied advertisers' first forays into radio in the 1920s. A medium prized for its potential for "cultural uplift," radio in the 1920s was revered for its intimacy and redemptive possibilities.[75] According to the advertising historian Roland Marchand, there were widespread fears in the advertising and broadcast industries that radio advertising would create "public ill-will," "breed public resentment," and constitute a commercial intrusion "into the sanctity of the home."[76] What if radio advertising alienated listeners and caused them to reject both the radio and, by extension, all types of advertising?[77] It was this potential for audience backlash that made advertisers wary of radio. As Marchand puts it, "To sponsor a program as a public service was commendable, but a direct pitch for a product would debase the medium."[78] Hence, a consensus emerged during the 1920s, endorsed by secretary of commerce Herbert Hoover, that supported "indirect advertising" and "unobtrusive publicity that is accompanied by a direct service and engaging entertainment to the listener."[79]

Seventy years later, methods of navigating the new terrain of internet advertising borrowed similar tactics: unobtrusive publicity, indirect selling, and useful services characterized the "informative" approach to web advertising. Although this approach was typical in 1994 and early

1995, the social context and industrial logic that informed it are rarely explored. Many descriptions of these earliest web pages refer to a phenomenon known as "brochureware," a pejorative term used to denote websites with little interactivity; they were mere collections of scanned documents slapped on a web page.[80] But the informative approach played a crucial role for commercial organizations that were testing a new medium and trying to navigate an ill-defined territory between innovation and social norms. Go too far in one direction, and you risked offending a valuable market; go too far in the other, and you risked being lackluster. As the television scholar Amanda Lotz notes, it has become a truism among media-industry pundits that no one in Hollywood wants to be first, but everyone is lining up to be second.[81] From music genres to video-game titles, culture industries across the board balance the fear of failing when trying something new with the desire to be on the cutting edge when a new trend emerges.

For many of these early commercial sites, the informational approach was both the safest bet and the clearest acknowledgment that their sponsors were true pioneers invested in the newest technologies and cultural forms. With a focus on informing over selling, early commercial sites adhered to the prescription to provide a "service" and create something "useful" by utilizing multimedia and providing informational depth. The home page that Organic created for Club Med, for example, was explicitly unveiled in press releases as an "electronic brochure."[82] With around a hundred pages filled with pictures, sound, and QuickTime movies of different Club Med villages, however, this "brochure" was far from a digitized version of print materials. Rather, the multimedia components demonstrated the web's potential as a "glossy" interactive publishing platform.

A second model, however, ventured beyond the "informational" approach and borrowed codes from "cool" noncommercial sites to reconcile the values of internet culture with branded, sponsored, commercial content. "Cool" commercial sites subscribed to four basic strategies to make these sites "fun," "useful," and "participatory": They embraced serialized storytelling; they appealed to the values of gift economies; they provided some kind of service or information; and they strove to facilitate community by enabling participation, connectivity, and sharing. These guiding principles that informed the early logic of "cool quality," however, also meshed remarkably well with the goals of "relationship marketing," which had become a primary catchphrase for direct marketers by the mid-1990s.[83] Relationship marketing emphasized dialogue over monologue to create "one-to-one" relationships between marketers and individuals. Working within the codes of cool, marketers were able to gesture toward established Net culture while also aligning themselves with the future of "one-to-one marketing." The website for Zima is particularly exemplary in this regard.

Zima's "Tribe Z"

First introduced in select markets by Coors Brewing Company in 1992, the clear malt beverage Zima was nationally rolled out in early 1994 and became faddishly popular for a short while before becoming a national punch line a mere two years later. In 1994, though, Zima's marketing strategy was being hailed as a unique, innovative, and tech-savvy way to reach a young, hip, mostly male, Generation X audience. As one of the original *HotWired* sponsors, Coors launched a website for Zima in late 1994 in tandem with its banner ad placement on *HotWired* and its offline campaign to drive traffic to the site by printing "zima.com" on drink labels. For the majority of Americans in 1994, this message would have had little meaning. But for a small community of young internet users, the label sent a clear message that Zima was on the web. According to Zima's brand manager, using the internet was "the perfect matchup with the type of people we were going after, people who are open to new things and who are leading their peer groups in using new technology."[84]

To produce the site, Coors hired the interactive agency Modem Media (based in Norwalk, Connecticut) rather than going through its traditional advertising agency, Foote, Cone, and Belding. Unlike the web service providers that sprang up to offer web development and hosting services after the launch of Mosaic, Modem Media got its start in 1987 as an "electronic direct marketing agency," creating electronic shopping malls for commercial online services, 1-800-number telephone response marketing campaigns, fax-on-demand services, and interactive media for CD-ROMs and kiosks. Modem's idea for Zima was to build an "on-line community" called Tribe Z that was packed with content: games, downloadable images and sound effects, regional bar directories, link

Figure 2.7. Tribe Z section of Zima website. Fahey, *Web Publisher's Design Guide for Windows*.

directories, communal spaces to congregate with other users, and an ongoing serial narrative starring an unseen character named Duncan, a tech-savvy, twenty-something Gen X web surfer.

The main navigational metaphor for the site was an image map of an interactive refrigerator ("the fridge") that emitted a "creak" sound when clicked on. Viewers could click within the fridge graphic to access various sections of the site. "Views" (Zima images), "Icons" (desktop icons), "Earwacks" (short sound clips), and "Diversions" (shareware games) all constituted "free cool stuff," downloadable multimedia files that situated Zima within the internet's gift-economy culture. "This was when getting stuff for free online was a big part of the draw of the internet," senior copywriter Charles Marrelli told me. "Little icons you could use in place of the default graphics on your desktop—people were into customizing their computers, and these graphics were like little refrigerator magnets you could play with. Part of it was about just seeing what your computer could do."[85]

More than just a destination for downloading "free cool stuff," the Zima site wasted no opportunity to demonstrate its thorough integration into web culture. Featured on an episode of PBS's *Frontline* in 1995, Modem Media co-founder G. M. O'Connell explains the site to the host, Robert Krulwich: "Zima is a place to go, to sort of find out what's going on with the internet, . . . a cool place, a place that's going to help you out, a place where you can see some of the best stuff that's happening on the internet."[86] The site aimed to accomplish this by looking for any opportunity to embed cool links to outside content. For example, to keep users coming back, the website featured biweekly serial installments that chronicled Duncan's adventures as he pursues a woman named Alexandria, who later becomes his girlfriend. According to Marrelli, Duncan was created as the "typical Zima drinker," built to "convey a sense of 'coolness' and 'regular guy' appeal for the Zima product."[87] The segments were short—usually a hundred words or less—and each featured an image, an icon, a sound, and a hyperlink to an external site on the web. For example, this is the episode for December 29, 1994:

1994 × 1995 RESOLUTION
Two minutes and counting.
    Jayson, party hat properly cocked, shirt appropriately untucked, gave Duncan a high five.
    "Isss gon be-nineyfive-man."
    Duncan tipped his Zima, "Yeah, dude . . . party on!"
I resolve to never take Jayson out in public again.
One Minute.
Gina's **noisemaker**
    "Duncaaaaan! Haappy New Yeer! Kiss me!"
I resolve to make sure Gina gets home safely.
10, 9, 8, 7 . . .
Out of the corner of his eye, Alexandria. Alone.
I resolve to ask Alexandria out.
The ball **dropped**.
In Two Weeks: Duncan calls his sister.[88]

In this text, the word "noisemaker" was a hyperlink to an external sound file, and the word "dropped" linked to a web page at the University of

Kansas called "URouLette," a random URL generator that would shuttle users to unknown destinations on the web. "We're actually helping the consumer surf the 'net," explained Mark Lee, Zima's brand manager. "We're trying to add some value by saying, 'Hey, let Zima show you some things we think are pretty cool.'"[89] Modem Media creative director Jim Davis likened Duncan to "a cursor or a navigational device to get the consumer to understand some critical things about the brand."[90] By integrating the outside web into Duncan's world, the Zima site was positioned not as a "gated community," like the private commercial online services, but as an embedded part of web culture.

Part of communicating that connection to web culture was establishing a particular tone or personality in the style of writing. Early commercial sites often featured a copywriter as the lead creative, since storytelling served a crucial role in low-bandwidth environments. One web-marketing guide noted, "In traditional publishing, copywriting is a separate component from design. With the advent of hypertext, however, the copy on your web site is not just writing, it is an interface."[91] In my conversation with Marrelli, he used a phrase associated with radio, "theater of the mind," to describe the work of bringing Zima to life through Duncan's stories and the collections of links that Zima provided to other cool sites online. Link collections were presented as a form of "giving back" to the Net community—whether in playful ways, such as following Duncan's adventures through portals to other destinations, or as a useful resource, such as the regional guides to bars and nightclubs listed in Z-Spots.

However, to really take advantage of the Zima online experience, users would have to sign up to become registered members of Tribe Z, an affinity club for like-minded visitors who were given access to one another as well as even more free stuff. After users signed up for Tribe Z by submitting their names and email addresses, they were given a unique password (allowing "unobtrusive tracking" of users' activity on the site) that could be used to access an email directory of other members and entrance to "the Freezer," an exclusive section of the Zima site with "cool digital gifts" and the possibility of winning T-shirts, posters, and caps. The Freezer section was framed by Modem staffers as "an experiment in participatory design." Using the visual metaphor of an unfinished loft, the members of Tribe Z were invited to participate collectively in the loft's design (see figure 2.7), a strategy designed to help consumers derive a sense of ownership, community, and involvement in the site.[92] The long-term strategy was to eventually allow Tribe Z members to "move in" and place their own home pages within the Freezer, thereby offering the brand as a personal haven that consumers could inhabit. "The Tribe Z member then becomes a cohort with the brand, a contributor of content, in the online advertising campaign."[93] If nonmembers of Tribe Z tried to access the additional content in the Freezer, they were told they were "not cool enough yet" and were referred to the area within the Zima site where they could join the tribe.[94] As the site evolved, new content and sections were constantly added, and viewers were often encouraged to share their own contributions by uploading images or games to "The Bin" and adding comments through a section called "Graffiti."

The Zima site was one of the first commercial web destinations that integrated online and offline marketing tactics using storytelling, gift economies, services, and community-building tactics that went far beyond the safe brochureware strategy that was common at the time. The project helped to bolster the roster of Modem Media's clients, moving the interactive agency closer to the powerhouse that it was soon to become. Just months after the Zima site launched, Modem beat out advertising agencies McCann-Erickson Worldwide and Young & Rubicam to become AT&T's interactive agency of record, a new designation that became popular in early 1995 as marketers moved to step up their interactive involvement.[95] By the fall of 1996, the seventh-largest advertising agency in the world, True North, combined its TN Technologies unit with Modem Media, creating the largest online advertising agency in the industry at the time.

Ragú's "Mama's Cucina"

Ragú's website, Mama's Cucina (www.eat.com), was a tongue-in-cheek home page launched in March 1995. Van den Bergh Foods, the company that owned the Ragú sauce brand, bypassed its advertising agency, J. Walter Thompson, and instead hired the Ann Arbor, Michigan, interactive company Fry Multimedia. Fry, along with the freelance copywriter Tom Cunniff and a handful of Van den Bergh staffers, developed the site, which had an estimated price tag of around $35,000. The initiative to produce a website was largely attributed to the enthusiasm of a single worker. A September 1995 article in *Ad Week* explains, "Alicia Rockmore, an associate brand manager at Van den Bergh, got bitten by the web bug late last year and wanted to put up a site."[96] This practice was not unusual; many early commercial web initiatives were helped along by eager employees with a particular interest in the internet. Rockmore says the idea met little opposition at Ragú, especially when it became apparent that the website would be among the first for consumer packaged goods on the internet. By the time the site went live in March 1995, a spaghetti sauce on the World Wide Web was still considered a novelty, and the site gained substantial attention for that very reason. The site was featured in a number of web guides and "best of the Net" collections, including *1001 Really Cool Websites* (which filed the listing under the category "Special, Supercool Sites").[97] Between the March 1 launch and late August 1995, Mama's Cucina was covered by approximately fifty publications, ranging from the *New York Times* to the *Fresno Bee*.[98]

Going for an Italian-themed site that offered recipes using Ragú products, the web team came up with the idea of centering the theme on an Italian grandmother, "Mama," who would give Ragú a voice and a personality, one that was highly attuned to Net culture. "She'd be very sweet, but she'd also be very savvy about things like MOOs and MUDs, and she would just drop references to internet culture out of the blue," Cunniff, who wrote the voice of Mama, told me.[99]

Adhering to the graphical conventions of the early web, the front page featured a signature headline graphic with the site's title and an illustration of the fictional Mama, who welcomed users to her kitchen with a rotating series of phrases to give viewers the impression that there was always new content. These greetings ranged from the philosophical ("All problems in life get simpler after rigatoni and meatballs") to the intimate ("I think that new butcher, Willie, has a crush on me. He slipped me an extra veal chop last week for free").[100] Selections of these phrases were commonly reprinted in web reviews to highlight Mama's personality and demonstrate the quirky and distinctive attitude behind the site. Mama's tone is carried through every section of the site, framing all interaction with the Ragú brand through Mama's homey style. Later additions to the site included a gossipy Italian-themed soap opera, "As the Lasagna Bakes" (1997), which was so successful that it spawned a spin-off called

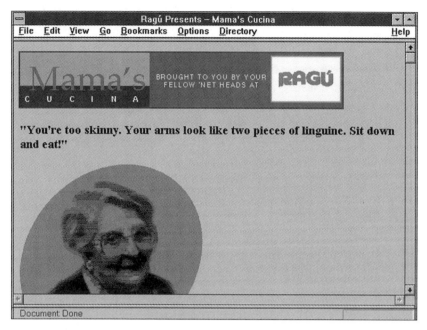

Figure 2.8. Mama's Cucina website for Ragú was launched in March 1995 at www.eat.com. The greeting at the top of the page constantly rotated so users had the impression of new content. This welcome reads, "Hey, how come you only drop by when you're hungry, eh?"

"One Love, One Linguini" (1998). Even the legal disclaimer is written in a folksy, family-style tone: "Mama's niece Ana, the lawyer, wrote this next part: Copyright 1995 Van den Bergh Foods Inc. All rights reserved." This narrative helped lend a sense of human appeal to the site, easing the tensions between commercial culture and the internet spirit of openness and sharing. To further ensconce the site into web culture, the signature graphic on the home page (see figure 2.8) announced that Mama's Cucina was "brought to you by your fellow 'Net heads at Ragú."

The appeal to the web's gift-economy mentality was also apparent in the blended content that was clearly Ragú inspired but included enough "fun stuff" that was not exclusively related to Ragú. Fry Multimedia, the site's developer, explains, "We focused on everything the brand touches, rather than the product itself. In other words, the site is about food, family, love and Italy."[101] Another section, called "Mama's Favorite Places," included a travelogue of Mama's favorite Italian cities, such as Milan,

Venice, Bologna, and Florence—each with population information, a description of the town, and a restaurant guide. There was a section on Italian art and architecture and a glossary of types of pasta. Visitors could enter contests for a free trip to Italy and play short audio clips of Italian phrases, courtesy of "Professore Antonio."[102] The audio clips also helped to boost the "multimedia appeal" of the site, since using the web to hear sound was still somewhat fascinating and worth visiting just for the experience. Users would click on the appropriate file type (AU, AIFF, or WAV) to play one of the Italian phrases, which included both the useful ("I would like a hotel room facing the Grand Canal, please") and the witty ("No, you're wrong. Pasta Toss has nothing to do with Italian football").

Perhaps most significantly, reviewers focused on the collection of recipes (raviolis, pastas, pizza, etc.) that were available in "Mama's Cookbook" ("I got a million delicious recipes for you. Scroll down—you'll find something you're crazy for. . . . Go, explore! Be the Marco Polo of the kitchen!"). Recipe sharing was a central component of Mama's Cucina, helping the site earn awards, positive feedback, and top honors in web guide directories. This food-centered focus on recipe swapping has a long tradition that dates back to UseNet groups in the late 1970s, a time when recipes were one of the few things that could be posted with impunity, as they escaped cross-border copyright laws. This put Mama's Cucina "very much in the spirit of the Internet—it does not just advertise; it tries to enhance the community feeling of the Net, rather than merely the fortunes of its creators," wrote one reviewer approvingly.[103] To further augment the community feeling, the site included numerous opportunities for user feedback and sharing. In addition to encouraging users to share recipes, Mama's Cucina urged users to contribute their own "stories around the family table" and to offer tips on Italian travel destinations and restaurants, which users earnestly did.

Perhaps nowhere was this communal, internet spirit more apparent than in the very domain name of the site: rather than associating Mama's Cucina with the commercial interests of Ragú or Van den Bergh Foods, the website was accessed through the communal and noncommercial-sounding domain name of www.eat.com.[104] The quality of "coolness" that Mama's Cucina tapped into was not one that emphasized visual sophistication; in fact, it was the relative lack of "graphical coolness" (in the first version) that added to the site's authenticity. By focusing instead on humor, character, community, and breadth of information, Mama's Cucina was able to integrate itself into established net culture.

Molson's "I Am Online"

As displacement transitioned to boom in the summer of 1995, excitement about the interactive future settled squarely on the World Wide Web, and marketers became even more invested in online initiatives. Financing huge multimedia productions to demonstrate their commitment to new media, advertisers aimed to take advantage of the ever-growing, young, well-defined, upscale audiences that congregated online. In the summer of 1994, there were few, if any, major marketers using the internet, but one year later, it was virtually a free-for-all, as various interests scrambled to stake a place online. As a journalist observed, "Summer 1995 saw the transformation of the internet frenzy from its first, subcultural, almost tribal stage to a national obsession."[105] Indeed, this notion of "national obsessions" was hardly a figure of speech. Despite the obvious global reach of the internet (including the World Wide Web's European roots at CERN and Tim Berners-Lee's desire to create a "universal" research tool), early commercial sites such as those for Zima and Ragú spoke to a clearly national audience of Americans. This was undoubtedly due to the geographic specificity of branding strategies designed to reach distinctive slices of populations, as well as the way ad accounts are conventionally assigned by national or metro market regions.

To serve the Canadian market, Molson Breweries launched an ambitious megasite called "I Am Online" (www.molson.com/Canadian/) in July 1995 to coincide with the beer company's relationship-building "I am" ad campaign for Molson Canadian that was running at the time. Conceived as a massive communications and cultural hub for Canadian internet users, "I Am Online" was a year in the making and had a price tag of over $1 million. The sponsorship model Molson adopted for the website was partly a response to established internet conventions, but Molson also had to work around federal laws restricting certain kinds of liquor advertising. Therefore, the site was imagined along the same lines as music events such as Molson Canadian Rocks, which the company sponsors. The MacLaren McCann ad executive Doug Walker explained, "It's something we want users to recognize as being 'brought to you by Molson—so enjoy.'"[106] This sponsorship model, in which the brand formed the backdrop to user communication, entertainment, and interactivity, helped Molson to avoid the feeling of a "corporate website" by acting more like an ISP that facilitated user-to-user interaction. "United States has America OnLine and CompuServe, but Canada generally had no coast-to-coast center for communication," explained a Molson representative.[107]

Unlike Ragú and Zima, however, the original idea for Molson Online came from Molson's advertising agency, MacLaren McCann Advertising. The Portland, Oregon, interactive specialist CyberSight was brought on board in early 1995 to produce the site and develop the communication tools that were to be a hallmark of the website. The visual identity of the home page took advantage of Netscape Navigator 1.1 extensions to create a look that broke with some of the conventions of the first era of websites. Although employing an image map to achieve more control over the placement of images was nothing new, the site managed to suggest an edgy, youthful, rock-inspired vibe by matching the graphic against a black background—a major difference from the drab, gray default screens of just a few months before (see figure 2.9). Of course, the black background would only be visible to viewers using the Netscape browser, but with Netscape's estimated market share of over 80 percent in the summer of 1995,[108] chances were good that Molson's target audience was viewing the website through Netscape. The content of the site focused on music, hockey, and Canadian cultural heritage, arranged within seven major areas so vast that the web team referred to each as a separate site housed within the larger megasite. Clicking on a link to "Soundbox" would take users to the Canadian Concert Calendar, where they could query a database to find musical events in their area, chat with musicians, read reviews of new releases, and download audio and video clips of bands. "The Word" offered "a compendium of provocative news" along with odd facts, humor, and cultural commentary derived from user submissions. A section called "Canadian Culture" provided movie, theater, and special-event listings. The most popular area, "Hockey Net in Canada," provided daily stats, a Hockey hall of fame, and a nationwide fantasy-hockey pool that was officially sanctioned by the

Figure 2.9. Molson's "I Am Online" website (1995) included seven main sections ("Pub Chat," "The Cooler," "Soundbox," "Hockey Net in Canada," "The Word," "Canadian Culture," and "Outer Limits"), presented in a graphical image map.

National Hockey League, which provided Molson with daily statistics during hockey season. Through the pool, visitors could draft a team of players from the NHL and play against one another, with a grand-prize trip to the Stanley Cup playoff game for the winner.[109] The site made sure to blend Canadian cultural content with internet cultural conventions by including a section of downloadable "cool stuff"—shareware, games, and screensavers—under the heading "Cooler," along with links to "other cool sites" on the web, which were collected under the title "Outer Limits."[110] Incorporating the gift-economy ethos and a willingness to point users to outside, non-Molson web content was framed as a way for Molson to deliver "corporate goodwill" to consumers.[111]

To facilitate personal communication, community, and national identity—the emotional center of the "I am" campaign—the website included numerous discussion groups as well as a section called "Pub Chat," in which users could explore a "virtual" map of Canada and chat in real-time with other users through the cross-Canada "MUSH" ("multi-user shared hallucination"). In addition, a webmail function was added to enable correspondence with personal post-office boxes that collected messages and notified users of new posts when they logged in, while a search engine enabled users to scan member profiles for others with shared interests. The entire megasite consistently emphasized interactivity and user services to provide an interactive center, rather than a typical commercial brochureware site on the web. Doug Walker from MacLaren McCann noted, "There are a lot of companies going online but time and time again they're just trying to create an online presence. You see a regurgitation of their annual report and a message from the CEO. That doesn't really work. What we're trying to create is a free online service. We've really tried to create a community people are going to feel comfortable in and are going to keep coming back to."[112] This approach, Molson's supervisor of internet projects admitted, was a response to "the Netiquette out there," a statement that registered Molson's concern that users would reject an overtly commercial message on the internet (as UseNet users did with the Canter and Siegel spam). Despite the substantial investment that a megasite such as Molson's required, a rationale was hardly needed. "It's bus fare," explained the Molson internet coordinator. "The amount we've spent on our website is less than half of 1 percent of our total marketing budget."[113] This paltry investment paid off handsomely—the site earned stellar reviews, attracted a large audience, and was still going strong years later, long after other "first-generation" websites had been gutted and revamped.

Conclusion

By closely examining the imprecise and ambiguous language that was used to describe and evaluate early web experiences ("useful," "best," "fun," etc.), this chapter has argued that the discourse of cool served an important need for early web users who were groping for the language to characterize a new structure of feeling. Although "cool" may seem as elusive and vague as "useful" or "fun," it actually played an important role in helping to give shape to a shared collective imagination that was being forged alongside the web. Displacement, if we recall, is the perception that something changes people's expectations about the future. In chapter 1, I suggested that there was no one single shock to the macroeconomic system, as economists often define this first stage of a speculative bubble. Instead, a particular conjunctural formation congealed around a shared image of the interactive future. Here I suggest that cool was the modality of displacement—it signals a new structure of feeling that was struggling for expression at the very edge of semantic availability. Cool links to cool sites, experienced as arbitrary juxtapositions that could collectively put the new spatial and temporal feeling of the whole web on display, offered a glimpse of the range of the web's possibility, which new users reported finding thrilling and strange.

Turning to how these "cool" qualities were borrowed and translated by commercial interests working to speak in the language of Net culture, we can see the important work that cool did for marketers hoping to establish a space for commercial web culture. Examining cool as both a new structure of feeling and a strategy mobilized to ease the anxieties surrounding commercial speech online, we see an emerging new media imagination forming alongside the commercial imperatives of capital.

Finally, this chapter has examined the social and participatory character of the early web and the important role that sharing, connecting, and communicating played. In chapter 3, we will see how a very different graphical design methodology came to signal quality web design as the dot-com boom gained momentum in 1995 and 1996. Only later did this format come to be known as the "read-only" web, in contrast to the more valued "read/write" qualities that came to be associated with Web 2.0 (see chapter 5). This oppositional framing of read-only versus read/write, however, has unfortunately become embedded in historical narratives of the web, wrongly suggesting that today's social, participatory, user-generated web only emerged after the bursting of the dot-com bubble. What gets obscured in that progress narrative is the coordinated effort involved in reconfiguring "cool" from a participatory, user-driven model to a readerly publishing paradigm. In other words, "read-only" did not result from naivety or technological limitations; rather, it was a hard-fought social achievement of a new group of players: media professionals.

3
Designing a Web of Legitimate Experts (1995–1997)

Yesterday, that great invertebrate in the White House signed into the law the Telecom "Reform" Act of 1996, while Tipper Gore took digital photographs of the proceedings to be included in a book called "24 Hours in Cyberspace." I had also been asked to participate in the creation of this book by writing something appropriate to the moment. Given the atrocity that this legislation would seek to inflict on the Net, I decided it was as good a time as any to dump some tea in the virtual harbor.
—John Perry Barlow, preface to *A Declaration of the Independence of Cyberspace*, 1996

It's been fashionable to attack what I wrote on the grounds that I thought cyberspace was sublimely removed from the physical world.
—John Perry Barlow, reflections on twentieth anniversary of *A Declaration of Independence of Cyberspace*, 2016

In 1996, John Perry Barlow penned *A Declaration of the Independence of Cyberspace*, a cyberlibertarian manifesto that famously proclaimed the electronic networks of cyberspace a zone of sovereignty, free from the terrestrial strictures of nation, government, or law. Conjuring the revolutionary spirit of the United States' Founding Fathers, the *Declaration* stakes a claim on this new virtual world on behalf of the explorer-colonists of cyberspace. "Governments of the Industrial World," it begins, "you weary giants of flesh and steel, I come from Cyberspace, the new home of Mind." Barlow's exhilarated prose celebrates a placeless place, separate from the confines of the material world, where the limits of embodied existence can be transcended "without privilege or prejudice accorded by race, economic power, military force, or station of birth" through a portal to another dimension: *cyberspace*.

Over twenty years later, the document reads like a relic from an ancient internet era, one premised on a series of dualisms between mind and body, virtual and real, ideas and matter, online and offline. *Declaration* is still routinely resurrected as a canonical example of the hyperbolic rhetoric, frontier metaphors, and digital utopianism that once figured the network as a vast, borderless, bodiless, meeting of Minds. In our current age of social media platforms with identity authentication systems that critically depend on real-name policies, it is easy to lose sight of how robust this vision of an autonomous cyberspace once was: it proliferated in popular media representations, figured in legal arguments and policy decisions, and guided early internet research activity. Across these domains, cyberspace was constructed as a new social realm apart from "real" life, which took place offline. Consequently, much of the early internet research was premised on the assumption that online sociality had an inherent cultural coherence that could be studied on its own terms—"in cyberspace"—divorced from existing social relations and practices.[1]

One consequence of the term's ubiquity in the 1990s is that the very idea of "cyberspace" has come to signify a historical break as well as a conceptual one. The "cyberculture" strand of internet research that once galvanized around theories of virtual community, identity play, and disembodiment is now seen as dated, as more people experience the internet as another part of everyday life.[2] "Social media," "algorithms," "platforms," "materiality," "reputation," "profiles": these are the terms that interest tech entrepreneurs and researchers making sense of today's internet ecosystem. As a result, "cyberspace" has become an anachronism; it gestures toward a network that no longer exists, embodying a postmodern perspective that feels insufficient in a post-dial-up, post-dot-com-bubble, post-9/11 social context in which internet use has become ordinary and bodies, borders, and the material world matter more than ever.

By the middle of the first decade of the twenty-first century, in the wake of these vast economic, political, and social crises, boundaries defining the meaning of cyberspace were being redrawn in a number of ways. In the field of internet studies, for example, attention shifted from the virtual as a site of disembodiment to the materiality of digital artifacts, evinced by new lines of inquiry in media archaeology, software studies, platform studies, critical code studies, and forensic materialism.[3] This "material turn" challenged long-held assumptions regarding digital media's ephemeral or immaterial nature through careful analysis of the working details, hardware, software, coding environments, and infrastructures within which technologies are made and used. As a corrective to earlier constructivist tendencies to downplay technology and critique "technological determinism," materialist approaches go "under the hood," as Jussi Parikka puts it, emphasizing medium specificity and the concrete operations and processual nature of computing over the analysis of representation and interpretation of meaning.[4] "Digital media are not simply representations but machines for generating representations," Noah Wardrip-Fruin explains. "Like model solar systems (which might embody a Copernican or geocentric perspective while still placing the sun and planets in similar locations), the operational and ideological commitments of digital media works and platforms are visible more in the structures that determine their movements than in the tracing of any particular series of states or outputs."[5] Viewed from the theoretical vantage point of contemporary media materialism or object-oriented ontology, the triumphant idealism of Barlow's mind-over-matter oratory (with proclamations such as "Ours is a world that is both everywhere and nowhere, but it is not where bodies live") only exacerbates the great divide separating the "old" ways of thinking about cyberspace and the "new" era of social media platforms.

The gap between "old" and "new" becomes even vaster when mapped alongside the dominant histories of the web that were set in place with the popularization of the term "Web 2.0" a few years after the crash of the dot-com bubble. This buzzword gained traction in Silicon Valley after a series of brainstorming sessions led by the technology publishing group O'Reilly Media, in an effort to "restore confidence in an industry that had lost its way after the dotcom bust."[6] Since then, concepts such as personalization, participation, collaboration, and sharing have been characterized as hallmarks for imagining the "next generation" web: Version 2.0.

This chapter challenges and relocates some of these techno-social, conceptual, and historical boundaries by using this 1996 legacy moment to tell a different kind of story about cyberspace. Rather than focusing on the myth of disembodiment or the rhetorical conditions of the

Declaration text itself, I foreground instead the backstory, the conditions and context in which its publication was initially sought, to return materiality and lived experience to the making of web histories. While Barlow's manifesto has become a famous piece of cyberspace history, the preface that accompanied it barely gets a mention. So let us begin with a practical question: what is this project that Tipper Gore and John Perry Barlow were working on—what was "24 Hours in Cyberspace"? By way of answering this question, this chapter examines how value and ideological conflicts inform the design of technological systems. I begin with the backstory of Barlow's *Declaration* to work through a complex of connected events that illuminate how human and nonhuman actors coordinate to accomplish mutual, overlapping goals while simultaneously obscuring underlying conflicts. In doing so, I build on Fred Turner's notion of "network forums," which describes how members of multiple, overlapping communities meet, generate ideas, and exchange particular "contact languages," memorable turns of phrase that help geographically dispersed communities cohere.[7] As discussed in chapter 1, "electronic frontier," "cyberspace," and "virtual community" were just a few of the notable terms that gained traction and meaning through their circulation through network forums such as the WELL (Whole Earth 'Lectronic Link) online bulletin board system and *Wired* magazine.

Turner's account provides necessary cultural and historical context that locates figures such as Barlow and *A Declaration of the Independence of Cyberspace* within a New Communalist countercultural spirit that can be traced back to Stewart Brand's Whole Earth Catalog from the 1960s. Through these network forums, Turner argues, scattered individuals come to experience a shared mind-set, even as they act from a mixture of motives and "speak from multiple registers simultaneously." So while network forums are sites where multiple groups come together into a single community, they are also importantly "economic heterarchies," complex networks with multiple and conflicting regimes of evaluation in operation at the same time. Heterarchies, according to sociologists Monique Girard and David Stark, are sites where competing value systems, organizing principles, and appraisal mechanisms coexist.[8] Journalists, activists, hippies, programmers, artists, and entrepreneurs came together in WELL forums and the pages of *Wired* to collectively forge a shared vision of the future under the banner of "cyberspace" and the

"New Economy" (which I address in more detail in chapter 4). This flexible interweaving of mixed motives, Girard and Stark suggest, is what distinguishes postindustrial forms of economic activity from the hierarchical organization that characterized systems of mass production. If hierarchies order groups into progressively smaller categories and subcategories, heterarchies are "complex adaptive systems" composed of multiple competing and coexisting logics and value systems that variously divide and unite groups as different perspectives emerge and recede from view.[9]

In this chapter, I focus less on the "shared mind-set" that emerges through network forums and instead draw attention to the underlying conflicts, contradictions, oversights, and political stakes that inform the design of the web but often get obscured in histories of cyberspace. Web design is about aesthetics, of course, but as this chapter shows, formal devices are not merely about beauty, style, or genre; aesthetic decisions are configured alongside a host of industrial, professional, technological, organizational, and economic considerations. By exploring the tangled mixture of motives, organizational models, and evaluative criteria, we may find ourselves better positioned to critically examine how myths such as "cyberspace," "Web 2.0," or "social media" cohere and become naturalized as "common sense."[10]

Interrogating Social Logics and Historical Boundaries

For a decade now, social media's participatory potential has become synonymous with Web 2.0 platforms such as Facebook and Twitter that facilitate the sharing of user-generated content. Even as the umbrella term "social media" supplants talk of Web 2.0 in the technology industry today,[11] the legacy of versions continues to hold sway as a periodization model in both popular culture and academic research that consistently figures social media as a twenty-first-century phenomenon. Matthew Allen refers to this type of web history as a "discourse of versions," which attempts to assert order and mastery over an anticipated technological future by claiming control of the meaning of the past.[12] Today, most popular accounts of web history follow a narrative in which the participatory read/write web evolved from what is now commonly remembered as the "read-only" quality of Web 1.0.[13]

Yet a closer examination of web history suggests that the boundaries separating the 1.0 and 2.0 eras distort our understandings of cultural and technological change. By bifurcating these periods, we lose sight of how the past is also characterized by complex overlaps, inconsistencies, and constant reconfigurations. For example, Michael Stevenson analyzes the design of one of the first commercial web publications, *HotWired*, between 1994 and 1997 and finds both parallels with twenty-first-century understandings of the participatory web and disjunctures that challenge easy distinctions between editorial-driven publishing and user-driven participation practices.[14] Even in O'Reilly's accounts of Web 2.0, we find hints of a more complicated genealogy that belies simple narratives of technological progress. In his second attempt to articulate the core values of Web 2.0, O'Reilly explains, "Ironically, Tim Berners-Lee's original Web 1.0 is one of the most 'Web 2.0' systems out there—it completely harnesses the power of user contribution, collective intelligence, and network effects. It was Web 1.5, the dotcom bubble, in which people tried to make the web into something else, that fought the internet, and lost."[15] This explanation both retains the chronological discourse of versions and simultaneously collapses these categories so Web 1.0 is one of the most 2.0 systems out there. Version 1.5, the dot-com bubble, is the outlier deserving explanation. If we follow O'Reilly, then, the question thus becomes not why the web became participatory or social in the 2000s but, instead, how the user-generated social web of the 1990s turned into "something else"—static, read-only web pages—a model that "fought the internet" by neglecting its true participatory nature. What this line of inquiry risks losing sight of, however, is that neither "read-only" nor "social" were natural qualities of the web: rather, both were achievements that took imagination and effort. Keeping this in mind, the task of examining how and why a read-only paradigm gained traction in the mid-1990s, only to be reconfigured in both imaginative and technical terms as a "social media revolution" a few years later, challenges the most common assumptions about social media platforms today.

Furthermore, this task presents an opportunity to rethink the dominant narrative of web historiography by considering how notions of "social" and "publics" have been historically imagined in relation to internet culture and figured in the design of media systems and platforms. By attending to historical genealogies of social media and read-only publishing, we are better equipped to understand how contemporary social media platforms install and naturalize a particular version of the social that depends on the algorithmic tracking of networks of association.[16] Lost in the dominant narrative of a shift from the 1.0 publishing models to the 2.0 user-generated social web are the ways these concepts are continually reconfigured through design and production practices, cultural and technological frameworks, institutional arrangements, and professional affiliations that yield both fruitful collaborations and conflicting visions. At the heart of these negotiations reside many of the classic tensions between publics and audiences that have long vexed media studies scholars and continue to impact discourses about mediated and networked publics.[17]

Configuration as Method Assemblage

To consider how the early web's promise as a new global communication network invoked a tension between how publics were imagined and addressed online, I trace the development, launch, and afterlife of two commercially sponsored web projects produced at the very start of the dot-com boom: *A Day in the Life of Cyberspace* (henceforth *1010*), launched on October 10, 1995, by the MIT Media Lab, and *24 Hours in Cyberspace* (henceforth *24 Hours*), launched on February 8, 1996, by the photographer Rick Smolan's production company, Against All Odds. These two projects represent conflicting visions of the future of the web: what it should look like, how users might participate, who is most qualified to design it, and what kind of logics, assumptions, and values ought to guide these efforts. These websites also exemplify two different ways of understanding and addressing a public: in *1010*, the web was treated as a social space for public deliberation and user-to-user participation; in *24 Hours*, the web was treated as a vehicle for delivering a publication to an imagined audience. In contemporary parlance, *1010* articulates social media to networked publics,[18] and *24 Hours* articulates web publishing to mass-mediated publics.[19] *1010* involves participatory platforms that allow users to engage with one another directly by sharing opinions and generating content; *24 Hours* recalls the direct address toward an anonymous mass audience, mediated through print or broadcasting channels that do not register user activity. But as this chapter shows, these dichotomies fail to engage with more complex ways that people, texts, institutions, technologies, and economies interact, develop particular qualities and logics, and change over time. Focusing on how publics and sociality figured in early web design offers one way we might begin to grapple with these entanglements.

I suggest it is within the socioeconomic context of the early dot-com boom that we might understand the read-only charge of Web 1.0 (or Web 1.5 for O'Reilly) not as a failure of imagination or a technological limitation but as a strategic and hard-fought social achievement of a particular group of actors, namely, media professionals, staking a claim for the future of the rising web. But if we follow the story of these two projects into their afterlife—that is, considering how the various technologies and related discourses were reconfigured as solutions to other problems in a changing socioeconomic context—we find that many of the core components of the read-only web publishing logic are reconfigured as a "self-publishing" revolution in the celebrated Web 2.0 practice of blogging. Meanwhile, the very technologies designed in the mid-1990s to support a social web similar to how it is imagined today—with personalization features, user tracking, and rating mechanisms—hardly inspired a social media revolution; instead, these technical components were reconfigured in the service of e-commerce and content targeting for customer management.

Undoing the 1.0/2.0 divide exposes a vast realm of divisions made to seem natural, divisions that, as John Law argues, do not necessarily equip us to understand the "messy reality" of the world. This raises questions not just about periodization but also about the methods we use to impose order by fixing the past into a coherent narrative that makes sense. Law introduces the term "method assemblage" to account for the ways knowledge practices and methodologies enact a "bundle of ramifying relations" that generate insides (what counts as present and accounted for), outsides (the world "out there"), and otherness (all that disappears beyond the boundaries of the project).[20] Law takes the term "assemblage" from Deleuze and Guattari to describe an ad hoc fitting together of heterogeneous components, not a stable arrangement but a "tentative and hesitant unfolding" in which "the elements put together are not fixed in shape . . . but are constructed at least in part as they are entangled together."[21] The concept of method assemblage is a call for the critical reinvention of more flexible boundaries that might better deal with the fluidity and multiplicity of the world.

Lucy Suchman offers the trope of "configuration" as a method assemblage that both attends to modes of production and offers a critical device for interrogating boundaries.[22] In the realm of computers, "configuration" quite literally refers to how a system is set up, the specific conjoining of hardware, software, components, and external devices that are put together for users to accomplish particular tasks. Similarly, when a team is assembled to produce a large or complex website, people bring learned lessons, former relationships, skills, and ideas, as well as code and technologies developed in other contexts. All of these pieces are constantly in motion, coming together, taking shape, only to be fit together in new ways and in different contexts, for different purposes and with different meanings later on. Configuration, Suchman proposes, is a device for studying how material artifacts and cultural imaginaries are joined together: "It alerts us to attend to the histories and encounters through which things are figured into meaningful existence, fixing them through reiteration but also always engaged in 'the perpetuity of coming to be' that characterizes the biographies of objects as well as subjects."[23] If we put aside the discourse of versions and approach web historiography as a site of ongoing configuration—"the perpetuity of coming to be"—we find that firm distinctions between read-only publishing and a read/write social web are difficult to sustain. As an analytical lens, (re)configuration offers a way to simultaneously engage the fitting together and the "tentative unfolding" of material artifacts (e.g., computer workstations outfitted with selected software), production practices (e.g., collaborative team-based models), and cultural imaginaries (e.g., how audiences/publics are imagined and addressed).

By examining *1010* and *24 Hours* in terms of configuration, I show how the social web's coming-to-be is not a linear progression toward participation but a negotiation of different models for understanding and addressing publics. As Michael Warner points out, the modern idea of a public refers to a social space of discourse. It is not a numerable collection of actual human beings but a social imaginary that comes into being by virtue of being addressed, and it is sustained through the temporal circulation of texts.[24] These mass-mediated publics associated with print and broadcast were premised on audiences as imagined communities.²⁵ With the rise of electronic communication networks such as the internet and the web in the 1990s and 2000s, new models were proposed: "virtual publics" formed through computer-mediated communication,²⁶ a "networked public sphere" that promised to disrupt the power dynamics of mass-media market economies,²⁷ and the "networked publics" of social media sites, where participation dynamics are shaped by platform affordances and complicated by the collapsing context that occurs when communication intended for a specific audience is received by multiple audiences instead.²⁸ With the internet, publics could be simultaneously imagined and numerable, sustained through the circulation of texts and shaped by the affordances of platforms. In the following sections, I consider how competing ideas about publics—as audiences to address and as users to engage—informed early web design practices along three intersecting vectors: discourses of quality and legitimacy, modes of address, and emerging temporalities associated with "real-time." I begin by framing *1010* and *24 Hours* within a set of conflicts about the "right" way to make the web at the very moment the web itself became a topic of public concern, the moment typically designated as the start of the dot-com boom.

Conflicting Visions: Imagining an Internet Spectacular

> The war of the coffee table books is brewing and it's being waged with the most unlikeliest of armies. On one side is a group of rumpled and angry scientists from MIT Media Lab, a world-renowned research facility known more for scholarship than imbroglio. On the other side is Rick Smolan, a celebrated photographer-cum-publisher who started "A Day in the Life" series of photographic essay books that swept the country in the 1980s.
>
> —Jon Auerbach, in the *Boston Globe*, January 18, 1996

The *Boston Globe*'s coverage of the "war of the coffee table books" has the elements for great drama: curmudgeonly scientists sequestered in high-tech labs versus a savvy celebrity photographer whose popular books adorn coffee tables across the United States. But like most wars, the terms of dispute are rarely the product of a single point of disagreement, let alone a contest over coffee-table books. On the one hand, this is a story of an ambitious web event that splintered into separate projects, conceived within the logics of two different institutional contexts. Both involved utopian visions of a digital networked future, a host of corporate sponsors that donated supporting technologies, and teams of experts and strategic partners charged with creating something truly spectacular to celebrate the global significance of the fledgling World Wide Web. On the other hand, this is a story of a bitter feud, representative of a clash between two different styles of imagining a digital networked future. By comparing how "quality" was imagined in these projects, we find different strategies for putting the power of the web on display as a new medium poised to address a global public.

In the spring of 1995, the MIT Media Lab, a renowned technology research center known for creative high-tech innovation, began brainstorming ideas for its ten-year anniversary celebration, which fell on the fortuitously digital date of October 10 (10/10). Imagining a worldwide digital event that leveraged the Media Lab's massive ATM (asynchronous transfer mode) fiber network, the concept for *1010* came together after a Media Lab graduate student, Judith Donath, suggested creating "a global portrait of cyberspace," a website composed of stories, sounds, and images contributed by users around the world.[29] This project was called *A Day in the Life of Cyberspace* (but known internally as *1010*), and from the beginning, the Media Lab framed it as a public event comprising the activity of users. Describing the challenges of planning the site, Donath explains that the goal was "to make it intriguing and enjoyable enough so that [users] would not only explore it, but actively participate in it." But the Media Lab was also mindful of quality participation: the level of engagement mattered, and the Media Lab specifically aimed "to create an atmosphere that fostered thoughtful contributions."[30]

To accomplish these goals and build interest over time, the Media Lab decided to launch the site as a ten-day countdown to the anniversary on October 10. Over a period of nine days, users would contribute content, either through email or through the website, in response to a different theme each day (i.e., privacy, expression, generations, wealth, faith, body, place, tongues, and environment). On the tenth day, there would be a big gala celebration, and a curated collection of contributions would be assembled "live from the Media Lab in Cambridge, Massachusetts," where "teams of professional editors and World Wide Web hackers working in 'mission control' at MIT" would collect, edit, and publish "the best of those bits" on the web.[31] These assembled "bits" would be preserved as a cyberspace "time capsule" from 1995 and later published as a book to be called *Bits of Life: Reflections on the Digital Renaissance*.[32]

Across these various forms—website, book, media event, anniversary party—we encounter a blend of values and concepts that are crucially linked to ideas about how a digital public would best be served by the web. *1010* was, on the one hand, a project that celebrated the digital as much as it did the Media Lab. Nicholas Negroponte, the Media Lab's co-founder and director, had just published his book *Being Digital*, and the ten-day launch of *A Day in the Life of Cyberspace* (www.1010.org) culminated with a gala symposium that began at precisely ten minutes past ten o'clock on the tenth day of the tenth month of 1995.[33] The project was also a celebration of user participation—but powered by the might of technological luminaries and the organizational capacities of an institute such as MIT. The call for participation framed this relationship as collaboration: "Your bits will become part of a global, public, community event—a canvas that we all paint together." This vision of the web—participatory, social, crowd-sourced, shared, and underwritten by a powerful platform provider—is typically understood as a 2.0 model; yet here we see these ideals expressed in one of the first attempts to celebrate the power and reach of digital global networks. The Media Lab was well aware of the limits of a digital time capsule to serve a future public: "Pages in the web are like footsteps on the beach. In a few years the bits and even much of the architecture of the web today may be rinsed away by wave after wave of new software."[34] Hence, it needed a book.

To produce the book, Negroponte brought the photographer Rick Smolan, known for his *Day in the Life* book series, on as a consultant in May 1995. But things hit a snag that August after a meeting with Kodak. Although Kodak sponsored each of Smolan's earlier *Day in the Life* books, this meeting did not go well.[35] As Kodak's senior vice president of marketing later explained, "If they intended to create a project that would offer value to a sponsor like Kodak, there wasn't going to be enough time to do the job right."[36] Shortly afterward, Smolan abruptly quit, a decision that he said boiled down to quality: "The more I talked to sponsors and technology people, the more advice I got to do it well or not at all."[37] Although the Media Lab's anniversary event went on as planned that October, it took place on a much smaller scale than initially imagined. For a lab that built a reputation on "personalization" technologies such as Personal Newspaper, Personal Television, and Conversation Desktop,[38] the task of integrating a mass-media model, in which "teams of professional editors" in "mission control" were charged with curating and assembling "the best of those bits," ultimately proved challenging for an understaffed editorial team. "The sheer volume of email flying off the printers made it impossible to read everything," explained Sasha Cavender, an editor who flew in to help. "While there were plenty of students who could write code, MIT didn't have the resources to edit, organize, and curate this material."[39] This was not an event designed to showcase the power of editorial oversight. Instead, the focus was on the anniversary celebration and the interactive capabilities of the site. Plans for the book were eventually scrapped.

This first conflict, then, illustrates that Smolan and the Media Lab understood "doing it well" in very different terms. For the Media Lab, the emphasis was on capturing the presence of users as they made sense of the internet in 1995 and designing technologies that could facilitate user-to-user interaction online. "It's not supposed to be the best of the Web," Donath told a reporter during the *1010* launch. "It's supposed to be a portrait of the electronic today—a frozen moment."[40] But to Smolan, a famed photographer known for taking on ambitious projects, the MIT event looked more like "an amateur photography contest" than the spectacle he had in mind.[41] After departing *1010*, he immediately began planning his own version, called *24 Hours in Cyberspace*, billed as "part event, part broadcasting, and part publication," which took place four months after the Media Lab's anniversary. *24 Hours* did not set out to celebrate the amateur. Instead, it was designed to be a global spectacle, carefully mediated by professionals and presented to the public as a tightly orchestrated "newsroom of the future."[42]

While the Media Lab urged the denizens of cyberspace to send it their "bits," the idea behind *24 Hours in Cyberspace* was to dispatch prize-winning photographers around the globe to capture the world in a day. These images would be transmitted back to the custom-built mission-control headquarters in San Francisco, where stories would be packaged on the fly by a team of eighty seasoned writers, editors, and designers and published as a visually stunning web magazine updated every thirty minutes around the clock on February 8, 1996. Aiming to show how the internet was transforming people's lives around the world, *24 Hours*, much like MIT's *1010*, offers a snapshot of the utopian ideologies surrounding visions of a digital, real-time, networked globe at the beginning of the dot-com boom. But *24 Hours* privileged media industries skilled at the professional packaging of the "world out there" for an audience to consume. Its version of "doing it well," in other words, drew on mass-media logic to frame the work of web production as the systematic coordination of creative talent and professional expertise.[43]

Needless to say, when people at the Media Lab read press releases announcing Smolan's new venture, tempers flared, and a public argument ensued across the media sphere. "They read like clones of what we had just done: a mission control, lots of editors, touting it as the largest Internet event to date," said one member of the lab. "It's the same story. And the total lack of acknowledgment really bothers us."[44] Smolan replied, "They wanted unedited bits. I wanted something crafted by professionals, that's what I do. They're scientists. . . . They're brilliant, but not publishing people."[45] While the projects' themes were indeed the "same story"—both evoked the world-in-a-day concept to create a time capsule commemorating the "digital renaissance"—the motivations for making the sites, the type of skills and expertise most valued, and the technical frameworks developed to create a "quality" online experience could not have been more different.

Significantly, these conflicting visions of what the web could be were unfolding at a crucial moment when public discourse about the internet was intensifying around a shared sense that "something important" was going on.[46] August 1995, when the partnership between Smolan and MIT unraveled, is the very moment typically designated as the start of the dot-com boom. This was the month that Netscape Communications, a two-year-old internet startup, held its initial public offering. It was a legendary debut that came to stand as a highly visible symbol of the internet's potential, mesmerizing investors and capturing the public imagination.[47] In a matter of one year, the web had gone from a niche computer network of techies to an aspiring mass medium, creating "a seemingly voracious demand for people who can create and manage Web operations."[48] *1010* and *24 Hours*, in important ways, represent the bids of different kinds of experts displaying contradictory styles of imagining the public performance of an internet spectacular at a moment when the meaning of the web was up for grabs. To examine how ideas about publics helped inform what "social" meant in these contexts, I consider the distinct ways each site addresses a public and how this address depends on two different ways of talking about real-time: as channeling social traffic and as automated publishing. Ultimately, I suggest, both were necessary components of what became known as Web 2.0.

A Day in the Life of Cyberspace: Configuring a Social Platform

If, as Warner suggests,[49] publics come into being through an address to indefinite strangers, what modes of address do we find in *1010*? First, and most obviously, there is the Media Lab's call for participation sent to listservs around the world that October: "We would like *you* to be part of the first global portrait of human life in the digital age." This personal direct address to an emphasized "*you*" speaks to the participatory nature of social media. (For instance, compare this to YouTube's invitation to "Broadcast Yourself.") Yet, as Paddy Scannell argues, this conversational direct address is a style of talk developed by broadcasters to speak to a mass audience not as anonymous multitudes but as individuals. He calls this a "for-anyone-as-someone" communicative structure, and it is fundamentally what makes broadcast *social media*. That is, in addressing "me" personally while I am simultaneously aware that countless others are also being addressed, broadcast talk offers the social glue that connects private individuals to the world at large. "For-anyone-as-someone structures in principle create the possibilities of, and in practice express, a public, shared, and sociable world-in-common between human beings."[50] *1010*'s call to participation, therefore, is not far afield from a mass-media address. But there is another address that positions users of the website, and it emerged from an explicit attempt to make the web feel more social by creating a sense of co-presence.

"Wandering the web is usually a solitary experience—one has very little sense of the presence of others," notes Donath in her documentation of the project.[51] *1010* employed a number of strategies to counter this tendency. The site was organized around the idea of personal time capsules—automatically generated profile pages that served as a hub for displaying site activity and communicating with other participants. It deployed a

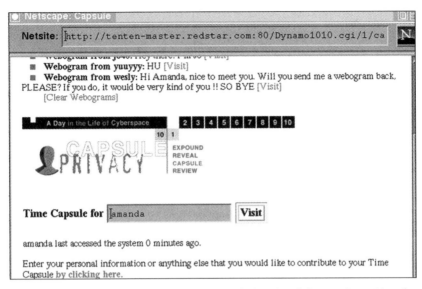

Figure 3.1. Screenshot from a personal time capsule (1995). Judith Donath / MIT Media Lab / Art Technology Group.

registration system whereby each user, upon logging in, was assigned a unique identifier that tracked his or her clickstream activity during that session. Most notable, however, is how these capsules were automatically updated as a consequence of participating in the site. If a user took a mini-survey in the "Reveal" section or contributed a comment about the daily theme in "Expound," those contributions would also appear in the user's personal capsule (see figure 3.1). The site therefore operated according to a very different logic of participation than do personal home pages, which were popular at the time. As Donath notes, although "ordinary home pages are designed once and then updated infrequently," capsules were automatically updated through a user's participation in the site.[52] Capsules therefore demonstrated a different temporality and mode of address. Because capsules displayed the unique activity of individual users and addressed users directly by name (e.g., "Time Capsule for Megan"), they involved a personal address to someone in particular. My capsule addresses me personally by name and is therefore not also directed to indefinite strangers, a prerequisite that Warner argues is necessary for an address to be *public*. How, then, did *1010* imagine and address a public? How was this built into the site's architecture?

Although every user was addressed individually by name, it is unlikely that each believed he or she was the only addressee. Instead, users understood that the site addressed indefinite strangers out there as unique too, each assigned her or his personal capsule. This communicative structure is crucial to the way twenty-first-century social media organizes and addresses publics. Attending to this reconfiguration requires accounting for more than how social media enables participation and self-expression; it also necessarily introduces a temporal arrangement in which users become attuned to "listening in"[53] for an algorithmic address, learning to read back the ways we are read by machines.

1010 combined this individual address with other features designed to help users "get a sense of the crowd" also online. Each page displayed a list of recent visitors and how long ago they were there (e.g., "2 people have visited this page in the last minute"). Users could therefore feel out other users through their temporal proximity and could communicate with one another with direct messages (called "web-o-grams") and indirectly by anonymously rating other users' discussion contributions. Algorithms used those ratings to manage the visibility of user contributions, so comments with repeated low ratings would slowly disappear. In two important ways, then, these design decisions simultaneously disrupt and reinforce modern ways of conceptualizing publics. Because the address alternates between indefinite strangers (an imagined space of discourse) and definite strangers (a numeral view of actual users), it troubles the distinction between discourse and dialogue, publics and bounded audiences, on which Warner's framework depends. But it also introduces a new temporal arrangement: by time-stamping participation, this mode of address creates not just a sense of public space but also the feelings of common time. At times used interchangeably with "live" or condemned as perpetuating the immediacy of an "eternal now," computing real-time is often associated with the elimination of a perceptible delay between processing and experience. But real-time is more perceptual, fabricated, and political than calculable. Esther Weltevrede, Anne Helmond, and Carolin Gerlitz suggest we think not of real-time but of "realtimeness" to consider how different platforms offer distinct forms of this temporality for specific users.[54] In the case of *1010*, realtimeness unfolds through the daily coordinated discussion around certain themes but also by time-stamping interactivity so other users are aware of the co-presence of others. This is actually a crucial component in how the internet constitutes publics. "Once a Web site is up, it can be hard to tell how recently it was posted or revised, or how long it will continue to be posted," Warner observed in 2002.[55] Time-stamping introduces a citational field and lays the groundwork for the web to operate as a "concatenation of texts in time."[56]

To accomplish this technical functionality, *1010* relied on sophisticated backend technologies developed by Art Technology Group (ATG), a local technology consultancy that partnered with MIT. Co-founded by a Media Lab alumnus and staffed with MIT graduates, ATG saw in *1010* a chance to test out a custom internet application called Dynamo, which it was developing to meet the needs of its corporate clients who were demanding more sophisticated web services. A set of technologies for connecting backend applications to the web to extend the web's capabilities beyond standard hypertext pages, Dynamo could automatically generate custom web pages for thousands of users on the fly, in response to user activity. The Media Lab's documentation of the project describes this as a form of "virtual crowd control," which "helped channel and improve the quality of input."[57] Algorithms operated behind the scenes to track and channel traffic, while user feedback in turn altered the content on display.

In 1995, Dynamo created a personalized social web experience that seems familiar to our twenty-first-century social media lexicon: profile pages, direct messages, wall posts, mentions, comment rating, and user tracking. While obviously not the same type of experience we now associate with Twitter or Facebook, *1010* employed some of the strategies and tactics of "social media logic," which José van Dijck and Thomas Poell identify through four interdependent elements characterizing today's ecosystem of connective media: programmability, popularity, connectivity, and datafication. Social media logic, they contend, channels social traffic in response to real-time interaction patterns, preferences and "like" scores, socioeconomic imperatives, and measurable data.[58] *1010* offers an example of an earlier moment when the mechanisms of this real-time logic were being worked out. This was not the production model that caught on in the mid-1990s. But as we will see, the technologies that made this possible did not disappear; they were reconfigured to solve different problems when this type of social web failed to take off in the 1990s.

24 Hours in Cyberspace: Global Real-Time Mass Publishing

The idea of capturing a global view of the world under demanding real-time circumstances, all in a single day, has long held appeal as a way for mass media to foreground their power to make the whole world instantly visible. In an analysis of *Our World*, a live two-hour international "satellite spectacular" broadcast across twenty-four countries in 1967, Lisa Parks examines how satellites were used to construct a fantasy of global presence by exploiting liveness and articulating it to Western discourses of modernization. The program accomplished this, Parks argues, by incessantly "spotlighting the apparatus," employing stylistic flourishes that call attention to television's global mode of production and its technical infrastructure as a spectacle.[59] "Liveness" was pre-scripted, or "canned," yet presented as a spontaneous "globe-encircling now."[60] Nearly thirty years later, in an effort to represent web production as the natural domain of media professionals, *24 Hours in Cyberspace* used remarkably similar strategies. It offered the mode of address of a live media event, one that promised the experience of watching history being made.[61] This involved hypervisualizing the behind-the-scenes infrastructure and putting "liveness" on display as a function of real-time publishing.

On the big day, February 8, 1996, media outlets swarmed the specially constructed 6,000-square-foot command center in San Francisco, which reportedly resembled "a cross between a daily newspaper at deadline and Mission Control for a NASA space flight."[62] Every moment of this event made visible the behind-the-scenes work of producing the site and emphasized how a team of experts managed to "pull off such a feat" in a single day. TV crews were sent on location with photographers so television viewers could follow the whole production process: we witness the subjects being photographed and the resulting images arriving at Mission Control, where they are sorted by the traffic team and passed on to the editors grouped into six "storypods," while sound clips from interviews with photojournalists are edited in the audio room. In a half-hour *ABC Nightline* special, a few fleeting shots of the resulting website are pictured, but the coverage focuses almost exclusively on the process that made the system work. This "backstage" material not only dominated news coverage but was also prominently featured in the *24*

DESIGNING A WEB OF LEGITIMATE EXPERTS (1995–1997) | 115

Hours website, book, and CD-ROM (figure 3.2). A series of PDF files were included in the "How It Was Done" section of the website, which took great care to illustrate the immense coordination between professionals, tools, and infrastructure required to visually capture and produce the world in a day. These media materials included a map of the Mission Control floor plan, a catalog of available story templates, and an elaborate series of diagrams visualizing the scope and coordination the project required (figure 3.3). This method of inviting audiences behind the scenes is one way the project aimed to establish authority by separating media professionals from an audience of listeners and readers. Users were invited to "participate," but these contributions were subject to approval by a panel of judges. A "parallel effort" of student work was separated from professional content into a "student underground" area.

In addition to the apparatus, "liveness"—with its privileged connection to broadcast media—played a crucial role in demonstrating the future of publishing. Although the *24 Hours* team called attention to its liveness through frequent comparisons to television and radio, the proj-

Mission Control, located in San Francisco, looked like a cross between a newsroom at deadline and the control room at a NASA launch.

Figure 3.2. Behind-the-scenes images of Mission Control were included in the published coffee-table book and accompanying CD-ROM (1996). Rick Smolan / Against All Odds Productions.

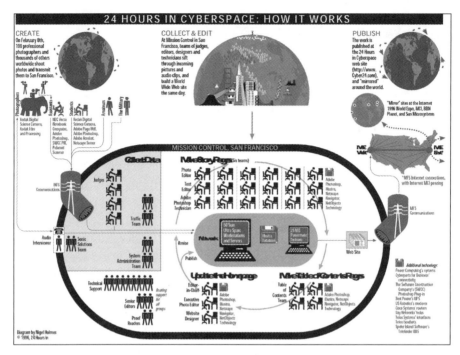

Figure 3.3. Diagram depicting organizational flow of *24 Hours in Cyberspace* (1996). Diagram by Nigel Holmes / Against All Odds Productions.

ect was never really "live," in the sense of transmitting images or sounds to audiences as they occur. Although updates were scheduled for every thirty minutes, deadlines were missed, and technical troubles caused a delayed start. Press releases announcing *24 Hours* pitched the event as "real-time photojournalism on a global scale," but the newsroom of the future was never concerned with responding in real-time to world events as they unfolded. Stories were researched and rough sketched well ahead of time, and the newsroom of the future was not willing or prepared to veer from the script. Most prominently, as the *24 Hours* event was taking place, President Clinton signed the Communications Decency Act into law, which caused a massive uproar across cyberspace. Although some coverage of the protests was later included in the archived site, *24 Hours* was criticized by groups such as the Electronic Frontier Foundation for not responding to cyberspace's biggest news of the day.[63] If the real-time of *24 Hours* was neither a way to respond to breaking news nor a live transmission feed from afar, what kind of realtimeness was it?

Unlike in *1010*, in *24 Hours*, real-time did not occur in response to user feedback on the "live web" but was instead an artifact of the production process—a content-management system that linked templates, a database, and a user-friendly graphical interface. This backend system could automatically generate HTML web pages at the click of a button, and it was this form of automation that inspired organizers to understand this event as "real-time photojournalism on a global scale." The production system was developed by a three-month-old software startup called NetObjects, formed by developers from Rae Technologies (an Apple spin-off that created information navigation applications), and Clement Mok Designs, Inc., an information design firm. "In most computing projects, the content is shaped by technological constraints," Mok explained. "On this project, however, it was the other way around— content was the force behind most of the decisions made about equipment, personnel, and the general dynamic of the project."[64] In other words, the type of content driving all other decisions presupposed the public as a reading audience, receivers of packaged professional content.

For media professionals to feel this domain was their own, the central role of HTML in web authoring had to be demoted. Those who were building the site needed to be able to plug content (images, headlines, captions, photos, stories, audio clips) into prebuilt templates that automated the technical work of coding websites. SitePublisher, the prototype that NetObjects designed for *24 Hours*, was described as "an automated Internet publishing system for 'non-techies'" that provides "clean, professional magazine layout and high-quality graphics."[65] That this high-quality, professional web magazine was produced in twenty-four hours by nontechnical users with barely three hours' training and no prior experience making web pages all served to demonstrate that the future of the web was not about hand coding HTML pages. It was instead dependent on the skills and talents of experienced writers, editors, and photographers—the content experts. Combined with other circulating web discourses that articulated "high quality" to visual styles associated with print aesthetics, this production model came to represent the codes of professionalism that many new websites—both commercial and noncommercial—aspired toward.[66]

The Social Life of Web Objects: A Market for Dynamo and Fusion

We could certainly end this story here. Both projects struggled with technical difficulties on the day of their events, and both, ultimately, were "successful." But if we follow these technologies into their commercial afterlife, we can see how they both became attached to new cultural imaginaries as they were reconfigured to address new problems in different contexts. It is this work of reconfiguration that alerts us to the difficulties and political stakes involved in drawing boundaries.

ATG released Dynamo as a commercial software package in November 1995, a month after the Media Lab's anniversary event. What made Dynamo particularly useful for web developers was its capacity for handling a library of applets called "Web Objects" (later renamed "Dynamo Gears")—independent, customizable, and reusable mini-applications. The Web Objects designed for *1010*, such as Session Tracking, Discussion Threads, Send Mail, Activity Monitor, People Watcher, and Live Links, could all be made available as a library of applets available for use in any website running Dynamo. Web Objects, in other words, are a material example of configuration as a type of assemblage. In the context of *1010*, these objects were used to experiment with sociable interfaces that might make online participation feel more vibrant and meaningful.[67] But for ATG's corporate clients, these same technologies were promising for their potential as targeted advertising vehicles.

Session Tracking, used in *1010* to collect and manage user access information for the personal capsules, also helped webmasters gather clickstream data and serve ads that change in response to user activity. In 1996, ATG released Dynamo Ad Station and Retail Station, developer kits with applications for profiling users, rotating ads, and managing retail sales. By 1998, amid escalating dot-com euphoria and investor interest in e-commerce startups, ATG had become a leading provider of "relationship commerce solutions," offering a suite of e-commerce and relationship marketing applications "designed specifically to tackle the management of online relationships by applying personalization to each and every user experience."[68] This helped make ATG a market leader in e-commerce, and an initial public offering (IPO) followed in 1999, making its founders multimillionaires. Two years later, ATG was ravished by the dot-com crash, shares plummeted, and both founders ultimately left the company. But ATG weathered the storm, staying focused on its commitment to personalization as a core business principle. A decade after the crash, ATG dominated the market for enterprise e-commerce platforms and was acquired for $1 billion by Oracle in 2011.

Meanwhile, after a stellar performance on February 8, 1996, NetObjects SitePublisher was revamped into a commercial web-authoring application called NetObjects Fusion, which was released to the market in the summer of 1996. Fusion aimed to make "web publishing as easy as desktop publishing" and explicitly hailed new audiences as desired web creators: "non-professionals" and "novices" looking for an easy way to quickly publish online and the "graphically-challenged," who wanted a no-fuss method for producing "professional quality" websites.[69] Introduced to the commercial market, NetObjects Fusion reimagined web authorship on several levels. As a WYSIWYG ("what you see is what you get") tool pitched as desktop publishing for the web, it opened the web-publishing market to a broader range of creators and therefore diluted some of the boundaries distinguishing media professionals, technical experts, and amateurs: all were framed as potential users of NetObjects Fusion.

The case of NetObjects shows how specific interpretations of "quality" production were built into the very tools and production logics that were in turn used to build the web. But once released into the marketplace, these discourses of professional-quality production were harnessed in different ways, particularly as a tool for bringing the power of publishing to the people. "The problem," observed one reviewer, "is that too many people, groups and organizations who have relevant information to put on-line are held hostage by the few who know how to do it.... The software industry is finally delivering programs that will put meaningful Web publishing tools in our hands."[70] As a product that made web production more accessible to a larger audience, the application was articulated to an emerging personal publishing revolution, one that promised to democratize the web through access to powerful, affordable web tools. At last, "anyone who wants to talk with the world will have the means to do so."[71]

In this way, an application built to service the needs of nontechnical media professionals, who envisioned clear distinctions between professional quality and amateur content and who regarded the future of online publishing as a way for media institutions to quickly produce content for public consumption, was reconfigured as part of a new assemblage joining content-management systems, ease of use, frequent site updates, and template models with a cultural imaginary that embraced the power of self-publishing. It is within this reconfigured context that the rise of blogging and the participatory discourses of the read/write Web 2.0 found fertile ground.

4

E-Commerce Euphoria and Auteurs of the New Economy (1998–2000)

"Content is king." This bit of conventional wisdom positioned the web early on as an advertising-supported medium, one that followed a long media lineage from the daily newspaper to cable television. Those who uttered it with such conviction did so because they had placed a particular bet on the internet's eventual winning business model. One early believer was Michael Wolff, who chronicles the rise and fall of his online publishing company, Wolff New Media, in his 1998 book *Burn Rate: How I Survived the Gold Rush Years on the Internet*. Tracking the breakneck pace of industry change and his struggle to find the funding that would help his company survive its burn rate (the total expenses that exceed revenue each month), Wolff observes that in the internet industry, there are East Coast "content people" and West Coast "technology people" engaged in "a death star battle ... for the soul of the Internet." These two camps represent not just different philosophies or expertise but different business models and, hence, ways of rationalizing losses and attracting venture capital. "Technology money follows different assumptions than content," Wolff explains. Technology money is about backing a winner; different entities and approaches duke it out for market share, which leads to market dominance. Losing money, even a lot of money, in the process is par for the course. Ultimately, the company that becomes the dominant player in this winner-take-all model then provides a historic return on investment to its investors.[1]

Content money, on the other hand, is not about market dominance but about finding and exploiting the hits—the *Lucy*s, the *Star Trek*s, the *Seinfeld*s of this new medium. Wolff New Media was positioning itself as the "gold standard" of internet content. Marketing partnerships, traditional advertising, direct sales, sponsorship, content syndication—all of these revenue streams, Wolff anticipates, would make his company a very valuable investment partner. After all, he reasons, "the Web had the *feel* of radio and television and cable broadcasting."[2] He assumes that on the web, as was the case with radio by the 1940s, users would settle into habits and favor specific content selections, which would help to make the web a wonderfully "predictable" world for advertisers. Although sudden "paradigm shifts" were constantly shaking up the industry, there was one assumption that most of the content people were not rethinking. Wolff describes the "unpleasant epiphany" that he came to in 1997: "What if the Web, the Internet, this whole thing, wasn't, well, media? Four years ago, at the dawn of cyberspace, virtually everybody not strictly in the technology business found themselves saying, 'Wow, a new communications technology! Must be media. *New* media!' There was a leap there that certainly seemed logical at the time, so logical that I don't recall anyone questioning it. But now, to me, it seemed glaringly untested. After all, the telephone was a communications technology that never became 'media.'"[3] Tired of begging for money, beaten down from failed negotiations, and exhausted from the pace of his publishing schedule, Wolff studies the server logs and comes to terms with the "incontrovertible truth" about the internet: people do not read the way they read print. Without a clear demand for his carefully edited, polished, and targeted content, Wolff finds himself without a business plan and without the motivation to get behind an internet that is fundamentally about something other than content. He explains: "In the cacophony of Web analysis and instant analysis and overnight paradigm shifts, people were now insisting that what worked—the only thing working with any consistency, with any predictability, with any prospect for long-term business growth and profitability—was transactions. Shopping. More and more, with greater and greater confidence and comfort, people were buying stuff on the Web."[4] The old mantra "content is king" had apparently lost its crown. Now the pundits were pronouncing "commerce is king," a change that not only signaled the rise of technology people but also impacted the structure of the industry, the nature of the mergers under way, the types of ventures more likely to attract capital, the direction of internet strategy, the skills most desired of new media talent, and the ways users were being conceived. It was, as Wolff put it, another major paradigm shift.

This chapter examines the conflicts between these two different visions of the web. On the one hand, there was the ad-supported content vision, which saw the web as media and therefore approached it as entertainment, a creative product, and the work of respected authors who could deliver engaging experiences. On the other hand, there was the transactional vision, which framed the web as a vehicle for electronic commerce and approached it as systems integration, as the product of talented programmers who could engineer technologies that deliver useful consumer-friendly services. To situate these industrial and professional conflicts within the socioeconomic context of the euphoria stage of the dot-com bubble, a period in which low interest rates and accommodative monetary polices helped create a volatile financial environment that supported the vertiginous rise of internet stocks, the first part of this chapter maps the discourse of the New Economy alongside financial speculation and internet industry maneuvers. By the time markets entered a period of euphoria, which lasted from late 1998 until the spring of 2000, the New Economy was constructed as a transactional discourse that was intimately connected to the promises of e-commerce.

Because I want to focus on some of the ways these material and discursive negotiations found their way into the production of cultural forms, I devote the second half of this chapter to a case study of the rising popularity of the industry-standard web animation software Flash. In this climate of heavy consolidation and in the move to support e-commerce and prepare for initial public offerings, I argue, freelance designers and small boutiques used their creative technical expertise in Flash to win back some of the power that had been ceded to the new global megaagencies being assembled to rule the business of e-consulting and web development. By positioning themselves not as designers but as auteurs with a unique vision, many creative practitioners saw in Flash the means to reimagine the craft of web design and intervene in the e-shopping discourse that was gaining momentum between 1998 and 2000. These designers used Flash to offer a different vision of the web: a quality website offered users a superior experience, a "rich media" experience that pushed the boundaries of expectations and expression.

By turning to Flash producers and Flash websites as an alternative mode of production that challenged the logic of e-commerce, I do not want to suggest these were the only two sides invested in making the web. One might just as well look at webcam operators, Java developers, web standards advocates, or the electronic literature community. I have chosen to concentrate on Flash sites and the Flash community because this focus productively complicates so many popular accounts of dot-com excess and irrational exuberance underlying the collective memory of the dot-com era. Sure, there was plenty of greed, egos, and shocking IPOs of immature companies peddling vaporware at the height of the bubble. But the problem with this simple characterization is that it fails to engage with rational and emotional conflicts—between art and commerce, content and technology, creating and engineering—that informed web design. The rise of Flash during the euphoria period represents a transformative moment of web practice in which presumptions over the meaning of the web and how it ought to look and work were in flux. It anticipates a sea change in new media cultural forms, marking a shift to a more visual web where new notions of creativity come into play. If the secret to amassing "great, quick fortunes [was] all about imagining the future," as Wolff insists,[5] then Flash offered a way for designers to help others visualize the web in new terms. Storytelling, abstract art, and sensory experience—this was the future that was proposed through Flash. Of course, Flash also illuminates the struggles for power, relevance, and reputation as much as a new semiotics of the interface. For these reasons, Flash provides a useful window through which to observe the interplay of industrial, socioeconomic, aesthetic, and technological change within the discourse of the New Economy.

Discourse of the New Economy

Analyzing the discourse of the New Economy means examining the pattern of talk, action, and interpretation that structured business models and monetary policy within a particular historical context. Discourse, Michel Foucault points out, both produces power and also serves as a system through which power circulates. Discourses are "practices that systematically form the objects of which they speak."[6] They are, therefore, not just talk but expressions with real material effects. Foucault later introduced the term *dispositif* to expand the notion of discourse to include visual and spatial arrangements. He describes a "thoroughly heterogeneous ensemble consisting of discourses, institutions, architectural forms, regulatory decisions, laws, administrative measures, scientific statements, philosophical, moral and philanthropic propositions—in short, the said as much as the unsaid."[7] So discourse is not merely the collective output of talk; it is the system of relations that link language, ideology, and knowledge to the built world, its design, and its regulation. To analyze the discourse of the New Economy, then, is less about taking the buzzword itself as an object of study and more about concentrating on the nature of these connections.

As the new economy became the New Economy, particular ways of making sense of business, markets, users, and social life manifested themselves in economic policies, market valuations, website design, and technological systems. These practices and institutions structured and were structured by this particular, historical, discursive formation. New Economy discourse, then, was a product of postindustrial capitalism that formed a set of rules for talking about, acting on, and exerting power over markets, businesses, and the socioeconomic world—a practice that in turn fed the material conditions of its making at the close of the twentieth century. This discourse did not emerge fully formed in response to internet euphoria. Instead, the New Economy was shaped by ideas about the role of knowledge in the economy and was articulated in various instances to the service sector, to information technologies, to global networks, and finally, by the turn of the millennium, to electronic commerce. As different political, industrial, and economic actors aligned the New Economy to confirm and support their interests and platforms, the idea of an information revolution that was transforming the global economic system became manifest in economic and foreign policy, industrial organization, and investment strategy.

Although the discourse of a New Economy is often linked to the late-1990s boom in internet and high-tech stocks, it actually began to take shape decades earlier. The management theorist Peter Drucker, for example, coined the term "knowledge worker" in the late 1950s and was among the first to suggest an economic shift in which firms of the future would succeed not because of their ability to make products but because of their ability to generate and use knowledge effectively.[8] Daniel Bell developed these claims further in *The Coming of Post-Industrial Society* (1973), arguing that a number of institutional and functional changes were occurring in the United States that signaled a shift from a manufacturing-based to a knowledge-based service economy. Following the economic downturn of the 1970s and early 1980s, this shift from manufacturing to service in industries such as banking, research, education, restaurants, computers, and other high technologies—became known as the "new economy" (not yet capitalized) that politicians and business leaders around the world linked to future economic prosperity.

In the Reagan era, the new economy was constructed as an information revolution whose very existence was framed as yet another testimony to the crucial importance of free-market entrepreneurship and the government policies that give business free rein. For Ronald Reagan, the information revolution was a useful weapon in the fight against communism. In a speech to Moscow University students in 1988, Reagan evoked the promises of a new economy in connection with a Western spirit of openness and freedom. We are emerging from the economy of the industrial revolution, Reagan noted, an economy limited by the earth's physical resources. In the new economy, these limitations would be overcome as the human imagination triumphed over the material world, where "the most precious natural resource" is the "freedom to create."[9] By positioning high technology and knowledge work as imaginative, creative freethinking, Reagan masterfully connected the new economy to faith, freedom, and progress—a move that allowed him to forge common ground with his audience while also advocating basic ideological principles underlying liberal democracy, such as free expression and free markets. He proclaimed: "In the new economy, human invention increasingly makes physical resources obsolete. We are breaking through the material conditions of existence to a world where man creates his own destiny. Even as we explore the most advanced reaches of science, we are returning to the age-old wisdom of our culture, a wisdom contained in the book of the Genesis in the Bible. In the beginning was the spirit, and it was from this spirit that the material abundance of creation issued forth. But progress is not foreordained. The key is freedom—freedom of thought, freedom of information, freedom of communication."[10] As both threat and promise, the new economy of the Reagan era was a malleable resource for talking about the future in both economic and political terms.

A few years later, as a way to restore the nation's confidence following the 1990–91 economic recession, presidential candidate Bill Clinton harnessed the term "new economy" to his economic strategy and promise of a "new covenant." "The rise of the new economy coincides with the triumph of democracy and the end of the Cold War," he told an audience at the Wharton School of Business during the 1992 primary season. "The old economy of a generation ago rewarded countries whose firms had strong organizational hierarchies and strict work rules. In the new economy, our prosperity will depend instead on the capacities of our workers and our firms to change."[11] Clinton's ideas for revitalizing the economy, restoring competitiveness, and boosting the nation's productivity centered on "rebuilding America" by creating new national infrastructures in a nationwide effort to generate the kind of high-tech, high-skilled, high-wage jobs that would be required in the twenty-first century. While economists, management theorists, futurists, and sociologists were already circulating the idea that advanced capitalism was changing and a new economy was emerging, the Clinton/Gore administration's promotion of a National Information Infrastructure (discussed in chapter 1) laid the groundwork for a connection between the new economy and the internet. By the summer of 1994, Clinton was referring to "global computer networks to exchange ideas and money" as the "nerve system" of the new economy.[12] Barely five years later, many popular commentators would forget that the new economy was ever about anything other than the internet.

The rising stock market in 1995 and 1996 helped revitalize new economy discourse as commentators looked for ways to explain market gains. However, the idea that the new economy was based on a transition from manufacturing to service sectors and knowledge work was quietly losing traction. Instead, the new economy was optimistically framed as a *new era* economy; the future was here, and it ushered in a new paradigm in which rapid technological change and global competition created remarkable economic growth. The economist Robert Shiller notes that stock-market expansions have often been associated with popular perceptions that the future is brighter or less uncertain than it was in the past—this was the case in 1901, the 1920s, and the 1950s–60s—and "new era thinking" accompanied each of these stock-market peaks in the United States.[13] Although conventional wisdom interprets stock-market gains as a reaction to new era theories, Shiller argues that the stock market actually *creates* new era theories, which emerge to explain the boom. "A stock market boom is a dramatic event that calls for an equally dramatic interpretation," he suggests. "Whenever the market reaches a new high, public speakers, writers, and other prominent people suddenly appear, armed with explanations for the apparent optimism seen in the market."[14] The new era discourse promoted by these experts sustains and amplifies the boom and serves as a feedback mechanism, which can create speculative bubbles. In this way, the discursive formation of the New Economy enacts the relations of power/knowledge that authorize the discourse as true and, in turn, produces the real material effects that produce the discourse.

The language of experts has a very strong influence on stock-market activity. Stock-market analysts pay close attention to the meetings of the Federal Open Market Committee (FOMC), the Federal Reserve committee that sets the target interest rate for federal funds. When the committee lowers interest rates, stock prices rally; when it raises interest rates, stock prices typically fall. For these reasons, Wall Street analyzes every word uttered by the chairman of the Federal Reserve, looking for hints as to whether a rate change is on the way. Since Fed administrators know this, key phrases are sometimes intentionally incorporated into speeches in the hope that market activity will adjust itself in anticipation of an upcoming change, which may be enough to stabilize the economy without actually increasing the rate.

One well-known example of such jawboning was Fed chair Alan Greenspan's use of the term "irrational exuberance" in his December 1996 speech titled "The Challenge of Central Banking in a Democratic Society" before the American Enterprise Institute, a research organization in Washington, DC.[15] Even though his lengthy address concerned the history of central banking in the United States and the role of the Federal Reserve in shaping monetary policy, the talk became famous for the one line he uttered near the end: "But how do we know when irrational exuberance has unduly escalated asset values, which then become subject to unexpected and prolonged contractions as they have in Japan over the past decade?" This rhetorical question immediately sent markets tumbling around the world as traders interpreted these remarks to mean that the Federal Reserve was considering raising interest rates because stock prices in the US market were overvalued. In this case, the market quickly bounced back, ending 1996 with an astounding 65 percent gain since early 1995. But the big question was this: to what did the market owe this incredible rise?

Business Week magazine is typically cited as one of the most aggressive proponents of the view that a new economy was reshaping the United States.[16] In a December 30, 1996, article, "Triumph of the New Economy," the magazine's chief economist, Michael Mandel, attributes the stock market's record growth "to the emergence of a New Economy, built on the foundation of global markets and the Information Revolution." He suggests that due to productivity growth, increased globalization, a boom in high technology, low inflation, and surging profits, the market's gains make sense. "The stock market's rise is an accurate reflection of the growing strength of the New Economy," he concludes.[17] By 1997, the discourse was now commonly being referred to as the "New Economy," the capital letters carrying the great weight of a paradigm shift. This New Economy was most enthusiastically adopted by certain journalists, investment analysts, and business media to justify the bull market. As the *Economist* reported, "A strange contagion is spreading across the land: the belief that technology and globalization promise unbounded prosperity and render old economic rules redundant has infected American managers, investors and politicians with remarkable speed."[18]

Claims that the "old rules" governing economics no longer applied came in both moderate and radical forms. But what were these old rules, and why did New Economy proponents believe that they no longer held? Economically speaking, the old rules involved the use of indicators to assess how well the economy was doing and how it would perform in the future. In particular, three key rules dominated macroeconomic theory.[19] First, most economists held that there were limits to how fast an economy could grow; the economy's "speed limit" was defined as the fastest economic growth rate that would not ignite inflation. In the late 1990s, the noninflationary speed limit was usually calculated at around 2 percent given the productivity rate at the time—exceed this, many economists believed, and the result will be higher inflation and an economic downturn. Second, it was believed that unemployment could not fall below a certain level without inflation rising. The negative correlation between the two is known as the Phillips curve, which posits a trade-off between economic growth and inflation. Finally, there was the understanding that an expansion in the money supply causes inflation. An increase in the amount of currency circulating decreases the value of money and hence causes prices to rise. But as the dot-com boom picked up steam, New Economy advocates at *Business Week* noted that the economic environment in 1996–97 was marked by a period of sustained economic growth averaging nearly 4 percent. This was almost double the speed limit; yet instead of rising, inflation was actually falling. "Something is going on that traditional economists can't explain," wrote Stephen Shepard in November 1997.[20]

From Boom to Euphoria

In the euphoria phase of a speculative bubble, traditional methods of establishing the value of a stock are ignored; conventional indicators such as earnings, revenues, cash flow, equity, and dividend yield—considerations that are typically employed to value the share of a publicly traded company—are disregarded as investors buy stock they know is overvalued, hoping to sell it to a "greater fool" for a profit. This age-old practice, known as the greater fool theory of investing, happens every time there is a speculative bubble, from the seventeenth-century "tulip mania" bubble in the Netherlands to the eighteenth-century South Sea bubble in England.[21]

According to the Minsky-Kindleberger model, expansion occurs when the availability of cheap credit makes borrowing easier, which opens the playing field to new investors who were previously shut out of the trading market. Couple that with the opportunities for profits that began with displacement and accelerated with boom, and soon the demand for the object of speculation presses against the supply. This results in a positive feedback loop, explains Kindleberger, since new investment increases income, which stimulates further investment and further increases in income—a process that leads to euphoria. When people see other people making money, they naturally want in on the game, regardless of how irrational the activity is or how little they may know about investing. As Kindleberger remarks, "there is nothing so disturbing to one's well-being and judgment as to see a friend get rich."[22] In the dot-com boom, the entrance of the everyday investor into the speculative game was further facilitated by the internet itself. Online trading sites such as E-Trade made buying and selling stocks a twenty-four-hour-a-day activity; investing chat sites and gossip boards further encouraged risk-taking and copycat behavior.[23]

The credit expansion that pushed the internet boom to a state of euphoria was preceded by events that started to unfold in the summer of 1998 when a South Asian currency crisis threatened to impact world economies. By mid-August, with news that the Russian ruble was in danger of being devalued, there was widespread concern that the economic malaise was broadening and could stunt the US economy.[24] Stock markets around the world tumbled in response to collapsing currencies in emerging markets. It was a turbulent time: Clinton had just publicly admitted his affair with Monica Lewinsky, Russia defaulted on its bonds and was on the brink of economic collapse, and there were strong fears that debt-laden Japanese banks would fail and thereby infect markets around the world. As the economic and political situation in Russia became more critical, the stock market continued to slump: the Dow lost 12 percent of its value in the last four trading days of August alone.[25] The threat of a world recession was suddenly very real: lenders stopped lending, and fears of a liquidity crisis were palpable. Then things got worse.

In mid-September 1998, Long-Term Capital Management (LTCM), one of Wall Street's largest hedge funds, announced impending bankruptcy after losing tens of billions of dollars in a matter of weeks. LTCM was highly leveraged and carried over $1 trillion in financial derivatives that intertwined it with every bank on Wall Street; its failure threatened to disrupt the global economic order in a catastrophic way.[26] "The Russian default turned out to be the iceberg for this financial Titanic," Greenspan later commented.[27] In response, the New York Federal Reserve Bank helped orchestrate an emergency bailout by a consortium of brokerage firms and banks—a highly unusual move given that hedge funds are virtually unregulated investments that speculate in high-risk trades in markets around the world. For these reasons, they are limited to "accredited" wealthy investors who do not need the same SEC protections as other financial instruments do.[28] But the ensuing rush to cash out of markets represented a broad panic that Greenspan feared would cause a devastating freeze. In response, between late September and mid-November, the FOMC dropped the federal funds rate three times in quick succession, pumping liquidity into the financial system and boosting the confidence of traders. While the immediate aim of the Fed's actions was to counteract the freezing up of financial markets, the rate cuts also sent investors a strong message. As Robert Brenner argues,

"The Fed's interest rate reductions marked a turning point not so much because the resulting fall in the cost of borrowing was all that great, but because it gave such a strong positive signal to investors that the Fed wanted stocks to rise in order to stabilize a domestic international economy that was careening toward crisis."[29]

From then on, commentators came to refer to the Fed's policy stance as the "Greenspan put" option.[30] In other words, investors now believed that Greenspan would intervene to prevent the markets from falling beyond a certain point. Indeed, as John Cassidy points out, when the market was climbing, Greenspan maintained a hands-off policy; only once did he raise interest rates since early 1995. But when prices started falling, Greenspan quickly changed course. "His reversal added to the growing belief that the Fed would always be there to bail out investors if anything went wrong," Cassidy argues, "and this made investors even more willing to take risks."[31] After Greenspan's third rate cut in November 1998, renewed optimism triggered a market turnaround so swift that the *Wall Street Journal* hailed it a "lightning reversal of autumn's doldrums."[32] By early October, the Dow had lost 20 percent of its value since its July peak, but six weeks later, the market soared past its summer high. Investors who had cashed out in fear of losing money in stocks were suddenly scrambling to get back into the market. As before the market scare, high-technology stocks were among the most popular investments. But the biggest gains by far were in internet stocks.

E-Commerce and the Transactional Discourse of the New Economy

The enthusiasm for internet companies fueled an "IPO craze" during the euphoria period that was largely led by e-commerce initiatives. Between October 1998 and March 2000, when the market peaked, there were 325 IPOs for internet companies; in comparison, during the boom period from August 1995 until October 1998, there were sixty-nine.[33] Furthermore, among the 325 new issues offered in the euphoria phase, the average first day returns were calculated at 91 percent.[34] The "stratospheric" and "gravity defying" valuation of companies such as TheGlobe. com, which achieved a first-day return of over 600 percent when it went public in November 1998, was breathlessly reported on CNN, CNBC, and CBS MarketWatch as a "feeding frenzy" attributed to individual investors who gobbled up anything with a ".com" name.[35] But it was prominent stock analysts and the business media that did more to authorize the discourse of the New Economy by constantly hailing a new set of logics that simultaneously acknowledged and rationalized the euphoria. Yes, the stock prices were through the roof, bull-market-inclined analysts duly noted in all of the well-circulated popular comparisons with Dutch "tulip mania."[36] But the internet represented a new frontier, and under these conditions, stocks of internet leaders may have been "cheap" because the market opportunities were so large.[37]

Although the New Economy discourse in the boom stage broadly linked stock-market gains to shifts caused by high technology and globalization, in the euphoria stage, this discourse was utilized to seal the connection between the internet and high stock valuations. E-commerce was the glue that brought these components together. All of the pieces of the New Economy puzzle—information technology, globalization, networks, knowledge, services, computers, and productivity gains—comfortably fit into the e-commerce business model. The "second wave of Internet IPOs," as CNBC characterized the growth of e-commerce offerings, was jump-started by the astronomical success of Amazon, which went public in spring 1997.[38] Like most dot-com stocks, Amazon had yet to turn a profit, but that did not dampen investors' enthusiasm. Amazon's founder, Jeff Bezos, openly acknowledged that he did not expect the company to achieve profitability for several years. Instead, the company concentrated on building market share and revenue with a "get big fast" philosophy that put brand name and an established customer base before earnings.[39] While e-commerce theoretically presented a viable business model, companies such as Amazon were losing millions of dollars a quarter. The new rules of the New Economy, therefore, needed new methods for evaluating the worth of an internet company that was doing quite well on the stock market but was not actually turning a profit.

In the old economy, the most popular way to quickly assess whether a particular stock was worth buying was to look at a stock's price-to-earning (P/E) ratio, which compares the relationship between share price and the company's earnings. But in the fall of 1997, the analyst Mary Meeker suggested that when it came to the internet, "we have entered a new valuation zone."[40] Portfolio managers desperately reworked numbers, trying to craft a rationale to justify their investment recommendations. "Trying to get your arms around the value of an internet stock is like trying to hug the air," a *Business Week* cover story acknowledged in December 1998.[41] Therefore, new methods for valuing internet stocks were invented, such as the Theoretical Earnings Multiple Analysis (TEMA) model, which relied on projected future earnings, or "cost per visitor," which estimated the value of website traffic.[42] With so little information about a company's long-term performance, valuation methodologies were based on marketplace potential and perceptions of future growth. These rationales served as yet another instance of how New Economy discourse formed a feedback loop that produced financial expertise, the systems for authorizing this expertise, and the models for evaluating market value.

Industry Consolidation and the Pursuit of the One-Stop Shop

One of the biggest challenges for aspiring dot-com companies that hoped to become e-commerce internet leaders were the low barriers to entry that characterized the web sector. This was, after all, the reason why so many companies could go from an idea to a public offering in twelve months at a multibillion-dollar valuation.[43] This was also the rationale behind Amazon's decision to plow all profits back into the business to benefit from economies of scale ("get big fast") and enhance its brand recognition—first-mover advantage would otherwise be lost as the playing field widened and e-commerce upstarts flooded the market. But e-commerce was by no means an inexpensive undertaking. There were technical infrastructures to custom build, backend systems to integrate, marketing plans to craft, and licensing deals to high-traffic portals to secure. Launching a website was now a major campaign requiring coordination, talent, creativity, and a lot of money. "When we decided to do this, and sat down with some venture capitalists, [they] asked us, 'Are you prepared to spend $40 million building a brand?'" explained the co-founder of an e-commerce site. "You really need to go the VC [venture capital] route to do this right."[44]

These capital-intensive projects that combined sophisticated technology and mass-marketing campaigns were a huge windfall for the internet developers, integrators, programmers, designers, and marketers who possessed the skills to turn these ideas into high-traffic websites. Between 1998 and 1999, the transactional discourse of the New Economy had a massive impact on the organization of web industries, modes of financing, and the makeup of players competing in the business to build the web. In the boom period, advertising agencies and interactive shops were the two main players that competed, combined, and collaborated in the struggle to rule the business of making websites. But the growth of e-commerce sparked a huge demand for internet systems-integration companies that could help corporate America integrate its old legacy applications with electronic commerce technologies. The business of creating a commercial website in the euphoria period was drastically different than it was just four years earlier. These sites were bigger, more sophisticated, and more expensive; they demanded bigger teams and more specialized skills to handle the demands of e-commerce. The founder of Organic Online explained in November 1998, "In 1994, we could get away with sketching simple brochure sites on paper and being awarded the business on the spot for $20,000. Today, that site would involve a blend of brand marketing, direct marketing, customer-service databases, content databases, competitive analysis, quantitative and qualitative modeling, supply-chain integration, banking interfaces, encryption, media plans and application software development.... Twenty thousand dollars would barely cover the travel and expenses of the pitch team."[45] In response, the web industry experienced a significant "shakeout" as companies consolidated to meet client demand, gain market share, and ultimately, become a "one-stop shop" that could handle all of a client's interactive needs.

"Get Big Fast": Consolidation of the Interactive Industry

As the interactive marketing sector started to mature by 1998, the sheer size of the companies that created websites and interactive marketing solutions grew at an astonishing rate. The merger of Modem Media and Poppe Tyson in early 1999 created the largest digital marketing organization in the world with operating offices on five continents in eight major cities across the globe. Meanwhile, the interactive agencies that managed to secure outside financing began to acquire other new media firms in desired markets to expand geographically, secure talent, build market share and revenue, and serve larger clients with bigger, more lucrative accounts. The most aggressive acquisitions were those of the interactive companies pursuing roll-up strategies. In a roll-up, dozens of far-flung smaller companies in closely related markets are bought and combined to create one large, industry-dominating company.

Take iXL, for example. After the Atlanta-based media entrepreneur Bert Ellis sold his group of television and radio stations in 1996, he formed iXL from the merger of a multimedia company and a video production house. He then transformed the company into a global internet holding company that aggressively bought up thirty-four web design shops, interactive entertainment companies, and interactive services businesses in less than three years. As an experienced media industries veteran who had already built and sold two broadcast companies for over $1 billion, Ellis was able to secure an initial investment of $37 million to fund iXL's aggressive acquisition strategy. "Bigger is better in this business," he explained. "It's harder to manage but easier to finance."[46] Indeed, finding investors was one of the things that Ellis did best. By steering the company toward the most lucrative aspects of the web development business, the rise of iXL during the euphoria period of the bubble provides a useful snapshot of how financing impacted the organization of the web industry, the pursuit of talent, and an industry-wide shift from a media to a technology focus.

By acquiring a number of web development shops, the company was able to move from a business based largely in video editing, CD-ROMs, and multimedia kiosk development to a more internet-focused organization. After a third round of equity financing in early 1998, iXL continued acquiring small and midsize web design and development shops in major cities, focusing on companies that served particular niches, such as building websites for the hospitality industry, the travel industry, or Hollywood entertainment. Several of these acquisitions were then consolidated to create large interactive agencies serving key geographical markets. For example, in May 1998, two more Los Angeles interactive agencies—Digital Planet and Spin Cycle Entertainment—were acquired and merged with other properties, BoxTop and Revolution Media, to create a new unit called iXL-Los Angeles, the largest office of iXL-West. At this point, the company billed itself as "a full-service interactive company specializing in the creation and management of digital content."[47]

But with this growth and consolidation came a noticeable shift in iXL's strategy and acquisition targets. "We've taken this integration opportunity to redefine how we service our clients' needs," said the vice president of client services at iXL-Los Angeles. Rather than focusing largely on interactive production as the company had in the past, iXL was now trying to attract more traditional clients, including those in automotive, banking, and healthcare, and so the company restructured its client services organization to add more business strategy and technology consulting experience.[48] These new functions were designed to help iXL evolve from a project-oriented company that built websites to a strategic business partner offering internet services and a long-term-relationship approach.[49] In short, iXL was moving from a web development firm to an e-services integrator, a company that built not websites but online storefronts and business strategies for some of the highest grossing companies in the world. "The real numbers are in selling stuff on the Web," Ellis realized.[50] To meet these needs, iXL purchased CommerceWAVE in the summer of 1998, an e-commerce solutions provider best known for creating online payment systems.

By December 1998, as internet euphoria was taking off, iXL was preparing for its initial public offering. Just one year earlier, the company was positioned as "a full-service video and multimedia production company," with stated core competencies in video production, web hosting, and interactive media.[51] But now iXL press releases framed the company as "a strategic Internet services company offering Internet-driven business solutions for Fortune 500 companies and similar international companies." Web design and multimedia production were replaced with new core competencies: "strategic opportunity assessment, Internet-based applications and systems development, information architecture and design, customer relationship management and the creation of 'rich media' environments."[52] By the time of the IPO, which came in June 1999, iXL had grown to 1,475 employees and was a key player in a new sector, the "Strategic Internet Services Industry." For "i-builders" and "e-integrators," as the new web development consultancies were known, the internet was no longer a new medium to orchestrate interactive advertising campaigns; it was a means by which firms could reengineer how they did business. Although many investors knew the outrageous stock valuations of dot-com companies were a bubble, the transactional discourse of the New Economy was also a very real, fundamental revision of the way global business works.

Technology + Strategy + Creative = The People Problem

As the case of iXL demonstrates, mergers and acquisitions were an efficient way to grow a company quickly. In order for companies to pitch the full-service solutions that clients were increasingly demanding for their e-commerce initiatives, specialists in three key areas were necessary: backend technical skills in internet systems integration, strategic consulting advice, and creative interactive marketing and branding solutions.[53] On a technical level, "internet integration" referred to the backend work involved in upgrading a client's computer systems so it could facilitate web transactions. Traditional systems integrators at this time—Andersen Consulting, IBM Global Services, Electronic Data Systems, and Cambridge Technology Partners—typically provided consulting expertise and IT solutions but did not specialize in e-commerce. This opened the door for a new category of interactive specialist that focused exclusively on internet and electronic commerce technologies and services. Internet integrators such as Sapient, Scient, and Viant brought technology and consulting to the internet: they determined how businesses could best use the web to reach customers and employees and then built the backend systems that enabled these applications. Needless to say, this kind of work was in very high demand. But as the e-commerce field became more saturated, making a transactional website work was not nearly enough. The desire to provide every component of a client's internet strategy drove a series of high-profile mergers and acquisitions for "e-business solutions providers" to become the one-stop shop that could provide the "magic trinity of strategy, creativity, and technology."[54]

Internet companies were expected to be unprofitable—taking a profit meant sacrificing growth and investment in the brand. Getting big fast and building a recognizable and trustworthy brand was thought to be the best way to become an industry leader. With the absence of profit, companies were valued by other means, namely, how much revenue passed through the business, how much future earnings might be expected, how competent the managers were, and how innovative the company was. But all of this obscures what many companies believed to be the number-one factor driving all of these mergers and acquisitions: the quest for talent. Businesses can be big in size and have many locations, but without the highly skilled people capable of building, inventing, designing, and integrating these "e-solutions," companies would fail to attract clients in the first place. With so many uncertainties surrounding the internet industry, talent is often the surest ground a business can stand on. As iXL chairman and CEO Bert Ellis explained, "If you're buying a television station, you're buying an awful lot of fixed assets and a lot of history and predictability." But in the internet business, it is much more difficult to assess value. "All we're buying is people." The goal behind iXL's acquisitions was to build a big team of talented people—and to build it fast. "This isn't a financial roll-up," iXL president Bill Nussey argued. "This is a talent roll-up."[55]

However, *keeping* talent after a merger was no easy task. Melding corporate cultures and retaining staff is a challenge even when bringing two creative studios together, let alone the cultural differences and workplace expectations that clash when creatives and computer engineers come together. When iXL acquired BoxTop, for example, the merger presented obvious benefits for both sides. BoxTop was a cash-strapped web design startup that built sites for movie studios and other entertainment companies, but it lacked technical expertise in building backend systems. iXL was flush with venture funding and looking to break into the West Coast market. But the reality of the merged workforces created significant tensions. According to Lisa Jazen Hendricks, a BoxTop cofounder who left a few months after the merger, BoxTop designers who were used to working at all hours were now expected to punch in and work traditional office hours.[56] Many designers were unable to stomach corporate culture and began to pack up—finding new work at smaller design shops or going freelance full-time.

Things got worse when iXL acquired Digital Planet and Spin Cycle. Here were essentially four different companies—BoxTop, Revolution Media, Digital Planet, and Spin Cycle—that were former archrivals, now expected to work together as one happy family. Egos clashed, different work processes caused confusion, and too many new technologies were introduced. It took six months for iXL to appoint a management team. People bickered over titles. All told, half of Digital Planet's original staff left, as did about 40 percent of BoxTop's original 140 employees, according to Hendricks and Digital Planet founder Josh Greer, who also left a few months after being bought by iXL.[57] BoxTop co-founder Kevin Wall stayed on and eventually became iXL's vice chairman. Creative types were bound to leave as the company focused more on corporate work and less on entertainment, he concedes.[58]

These conflicts complicate the practice of a "talent roll-up," or buying companies to accumulate skilled workers. Engineers and technologists are easier to assimilate into corporate culture and tend to thrive in managed project work environments.[59] But creatives in small companies that are acquired by much-larger companies often flee for more freedom, cooler projects, and less bureaucracy. People problems, then, become one of the fundamental glitches in the otherwise-reasonable idea of a one-stop shop. Bringing core competencies together in the service of diversification is nice in theory. But in practice, technology plus strategy plus creative has a hard time congealing because many creatives resist working in massive web integration companies.[60] So while industry consolidation meant more consulting expertise and backend support, there was a pressing concern that the mergers took a toll on creativity.

As the web became oversaturated with e-commerce startups and flocks of traditional retailers moving online, it became more important than ever to cut through the clutter. Although e-commerce was the biggest buzzword of the late 1990s, shopping carts and secure servers did little to convey the sexiness and excitement that new media evoked. Consequently, small, independent boutiques were highly sought after by internet integrators and transactional content providers eager to walk into pitch meetings alongside hip partners whose youthful, avant-garde creativity would offset the conservative image of corporate web consultancies. "We're walking in with suits; they're walking in with orange hair," said the global director for the e-commerce firm Ernst & Young. "The combination is powerful."[61]

There is, therefore, a counterpart to "get big fast," and that is to get very small and specialized. While small and midsized shops were gobbled up en masse by roll-up companies, advertising agencies, and growing interactive companies, there was still a great need for small design boutiques that could lend an air of edginess to a project. As Bernard Miège has noted, there is a permanent creativity crisis in the cultural industries, and producers must constantly be on the lookout for new forms or new talent.[62] As talent is linked to skill specialization in new media industries, many freelancers and small boutiques rely on their creative technical expertise in scripting languages, graphics, and multimedia software to maintain creative power in a consolidated market.

While creative web work necessitated the development and mastery of a range of technologies, one software package that became extraordinarily popular was the industry-standard web animation program Flash, distributed by the software company Macromedia. Using Flash to produce websites with motion graphics, sound, and sophisticated interactivity, small, niche boutiques and freelance designers were at the forefront of new approaches to web design that were vastly different from the static, silent, textual forms that imitated the aesthetics of print magazines or the functional commercialism of e-business solutions. The rest of this chapter presents a case study of the rise of Flash and its appeal among web designers during the euphoria period. Flash sites, I argue, helped designers and graphic artists present an alternative discourse to the e-commerce vision, one that emphasized creative expression and sensory experience over the transactional discourse of the New Economy.

The Rise of Flash

Although multimedia sites made with Flash seemed to burst onto the web with a sudden ferocity between 1999 and 2001, the software for authoring Flash content was first released in the summer of 1996 as a simple animation and drawing program called FutureSplash Animator, made by FutureWave Software. Just a few months earlier, Netscape Navigator 2.0 was released, providing a framework for third-party developers and vendors to "plug in" other applications into the browser to extend its capabilities beyond HTML.[63] The FutureSplash authoring program included a free Netscape plug-in that could be used to view FutureWave animation on the web. Not to be outdone, Internet Explorer 3.0 (IE) also added plug-in support, which meant that FutureSplash animation could be viewed through IE as well.

As the Browser Wars heated up, there were thousands of plug-in applications that could be downloaded for each browser to add "bells and whistles" to a website. Macromedia's Shockwave player, which enabled web support for interactive multimedia content authored in Director, was already being used to add animation to web pages. But since the software was originally developed for creating CD-ROMs, it was not optimized for the web and caused sites to run very slowly.[64] FutureSplash, on the other hand, was a much simpler program that could be used to create animation with very small file sizes. Because it relied on mathematical algorithms to create vector graphics, as opposed to bit-mapped images that literally map every pixel on the screen, animation could run smoothly and without a drain on processing power. When several interactive agencies began using FutureSplash to add motion graphics to prominent commercial sites, it gained a bigger user base and began to attract the attention of both designers and competing software companies that developed web design applications. Microsoft's MSN website used FutureSplash in an attempt to create what one producer described as "the most television-like experience possible on the Internet."[65] Fox deployed it in its website for *The Simpsons*, and Disney used it in its subscription-based online service, Disney's Daily Blast. FutureSplash garnered enough attention as a more streamlined alternative to Shockwave that Macromedia acquired FutureWave in January 1997 and relaunched the application as Macromedia Flash.[66]

Although the ability to add animation to the web was clearly an early draw of Flash, the software also solved a number of technical, social, and industrial problems that plagued designers working in the heat of the Browser Wars. In the competition between Netscape and Explorer, both companies were releasing browsers that incorporated their own proprietary tags known as "HTML extensions." Browser companies hoped these unique browser-specific features would seduce web creators into developing sites exclusively for their browser alone. Incompatibilities were further exacerbated with the different support each browser provided for generating DHTML (dynamic HTML) websites with multimedia elements. As Jeffrey Zeldman, co-founder of the Web Standards Project (WaSP) explains, "Things came to a head with the release of the 4.0 browsers in 1998. Suddenly Netscape and Microsoft were both trumpeting dynamic Web pages. No more static, print-like Web layouts! Images could move, buttons could beep, whole chunks of text could appear and disappear at the twitch of a hapless visitor's mouse. The problem, aside from obvious aesthetic abuses, was that Netscape and Microsoft had different and completely incompatible methods for creating dynamic sites. If it worked in Navigator, it would fail in Explorer, and vice versa."[67] These different implementations of code meant that it was not unusual for designers to create up to four different versions of the same website to accommodate users with different browsers and operating systems—a nuisance that vastly raised the cost of web development.

But when Macromedia signed licensing agreements with Netscape and Microsoft in 1998 ensuring that the Flash player came pre-installed in the Netscape browser and the Windows 98 operating system, most users no longer had to download the plug-in to view Flash content. For designers, this provided a way to escape the headaches of nonstandard HTML implementation. By producing websites entirely in Flash, designers could solve the technical problem of cross-browser compatibility by creating a single website that displayed uniformly across multiple browser versions. Furthermore, the vector files that Flash created made full-screen, high-resolution motion graphics possible for a fraction of the file size as a bitmapped image—which meant the technology could be used to efficiently deliver a rich media experience to modem users.[68] Finally, Flash was a software tool that made sense to graphic designers familiar with vector-based drawing software such as Freehand and Illustrator. Unlike HTML, Flash offered total control over typography and screen space.

But perhaps more than anything, Flash appealed to a whole new vision of the web, one that captured the imaginations of designers who were navigating the huge industrial shifts that were impacting the organizational structure of e-consultancies and the role of creative design in the euphoria period. Small creative boutiques and freelance designers saw Flash as one way to specialize their services, while also negotiating higher project fees at a time when the global powerhouses were so invested in e-commerce applications. Indeed, some of the first sites made entirely in Flash were created by freelance designers and independent studios that used their websites as a platform to showcase and refine their skill set. One of the first all-Flash sites to capture the attention of both professional and aspiring designers was a website called Gabocorp that appeared in late 1997 (figure. 4.1). Visitors to the site were greeted with a dramatic countdown segment complete with booming sound ef-

Figure 4.1. Screenshot from Gabocorp (1997).

fects: "Loading . . . 7: Sound Effects; 6 . . . 5: Background Sound; 4 . . . 3: Interface; 2 . . . 1 . . . almost done." Once the site was fully loaded, the opening text cinematically faded in, boldly proclaiming, "You are about to enter a new era in website design. This is the new standard for all things to come. Welcome to the new Gabocorp." Up until this point, most designers had never heard of Gabocorp before. But the site, with its spinning spheres, electronica background music, animated buttons, dramatic sound effects, and a graphical interface that rotated the full screen when a button was clicked, inspired an outpouring of imitators and parodies that proliferated for the next three years. Gabocorp, which gave the impression of an established web development firm, was actually the brainchild of seventeen-year-old Gabo Mendoza, an interactive developer who partnered with friends on paid web projects. This was part of the appeal of Flash: it offered designers the possibility to remake the web as well as their own professional and visual identities.

The new era that sites such as Gabocorp heralded was one that emphasized creativity, expression, and impact. Flash ushered in a new visual economy in which the most talked about and circulated websites were not necessarily the ones that facilitated e-commerce transactions but instead suggested new relationships between the user and the screen. As the Flash developer Colin Moock argues, Flash's popularity was based in its "hackability," its potential for modifying the system. Of course, graphic designers "hacked" HTML by using tags in nonstandard ways to gain more control over visual layout, but the type of hacking Moock refers to does not involve manipulating code. He explains, "I'm using the word 'hack' to mean 'a creative exploration, using any means possible (even ugly), that leads to some desirable (even exhilarating) result."[69] Flash's hackability, in other words, does not take place in the code; it hacked the meaning of the web through its visual expression on screen. Designers used it to modify or challenge conventional systems of representation online, which appealed to both advertisers and artists alike.

As dot-com saturation peaked, it became increasingly difficult for commercial sites to stand out from one another. Because Flash websites looked so different from the web that most users were accustomed to, clients began demanding Flash integration into commercial websites as a way to achieve "stickiness," a late-1990s buzzword that emphasized web content that compelled users to "stick around." Animation, sound, and interaction were believed to enhance the interface to the brand, providing "an experiential communication of a brand's core values." By 1999, Flash was high in demand. "There are headhunters and HR people whose first question is 'How are your Flash skills?'" one designer observed.[70] Interactive advertising and marketing agencies realized the tremendous potential this provided for engaging audiences with brands as Flash offered to turn the now-staid banner ad into a "rich media" experience. Small, niche design studios that specialized in Flash production, such as Juxt Interactive, Kioken, Hillman Curtis, Hi ReS!, volumeone, and 2Advanced Studios, became sought-after talent and were brought into projects by larger interactive companies and e-consultancies.

The rise of Flash from one of thousands of plug-ins to the industry-standard method for delivering animation and interactive multimedia online was not due simply to the technical superiority of vector animation or to the captivation of designers who imagined new creative modes of online representation. The technical and social dimensions of the software were cultivated by the strategic alliances Macromedia pursued and the company's tight integration of its suite of web design tools. As the multimedia programmer and consultant Bruce Epstein points out, "Macromedia cross-promotes their tools heavily to their loyal user base, just like Disney does with its movies and videos."[71] Because all of the software tools share common UI (user interface) elements, it becomes easier for users of Dreamweaver or Director to pick up Flash. With around 80 percent of professional web developers using Dreamweaver by 2000, this opened the door for a wider embrace of Flash.[72] While the proprietary nature of Flash helped ensure a well-funded marketing department to push the technology, Macromedia also released the Flash file format to the public, thereby inviting third-party tool creators to develop applications that could produce Flash files. This ensured that the Flash format was widely supported. Finally, the company developed a reputation for working closely with designers and developers, listening to their requests and adjusting the capabilities of the technology to address these needs.

Although the first version of Flash was a purely animation tool, each successive version moved the software deeper into a more robust application development environment. When Flash 4 was released in August 1999, a programming language called ActionScript was added, as was support for data collection that could be passed on to databases or other server software—capabilities that lent themselves to e-commerce applications. Version 5, released in the summer of 2000, took these programming capabilities even further so that Flash was able to effectively serve two markets: one for graphic designers interested in producing highly visual interactive experiences with sound and motion effects, and one for programmers/developers interested in creating games and dynamically generated content. Through these efforts, Macromedia helped turn Flash into a multimillion-dollar industry of not just software sales but numerous Flash-related products: training videos, classes, workshops, magazines, books for creative inspiration, books for learning code, and even books by design studios that deconstruct in painstaking detail the process of conceiving, building, and coding their websites.

Glitzy, well-publicized Flash conferences such as the semiannual Flash Forward Conference and Film Festival generated a huge amount of interest in the software and in the practice of web design. At the height of the dot-com bubble in 2000, there seemed to be no end to the demand for Flash skills and Flash code. Because the same new media production tools were used by both professionals and amateurs, the distinctions between hobbyist and creative professional were significantly blurred, an entanglement that impacted representations of expertise now that anyone could be a "designer." Peter Lunenfeld describes the "myth of Web design" as the enchanted discourse that was attached to the notion of the creative individual in the dot-com era, inspiring countless nonprofessionals to think of themselves as designers who were forging ahead on the cutting edge of technology and internet culture. "Something about the word 'design' . . . appealed to a huge array of people," Lunenfeld remarks.[73] And it was graphic design, in particular, that became the inspiration for so many self-nominated web designers to embrace design as a mass preoccupation. Noting how far removed the new discourse of web design was from the theory and discipline of graphic design, Lunenfeld laments that web design ended up appropriating the most commercialized discourses available, the "language of the Design Annual," and concerns for "the latest, the hippest, the newest, a language of the showcase and of promotion, rather than any kind of real critique."[74] Although it is true that design culture at the height of the bubble perpetually pointed to itself—to personal sites and portfolios and winners of Flash competitions—Flash embodied a struggle to resolve a set of contradictions that speak to this commercialism, attempting to gain power over it while simultaneously working within its systems.

Since it was skill development and mastery, not work experience or formal training, that played such a significant role in the growing Flash industry and its corresponding social world, many aspiring Flash designers were able to cross over from enthusiast to professional by building a personal showcase website to promote their skills in the absence of an actual portfolio. Showcase sites had the dual purpose of demonstrating skills and techniques to potential clients and employers while also catching the attention of peers to make a name for oneself in the Flash community. In addition to serving as a key venue for promotion, personal sites were treated as a playground or an exhibition space that provided an outlet for Flash users to probe the possibilities of the software, explore modes of creative expression, and share samples of source code that would otherwise not be accessible given the proprietary nature of the authoring software.

As designers built personal Flash sites to experiment with the technology outside the constraints of corporate projects, they pursued new forms of representation by investigating concepts such as friction, elasticity, magnetism, and generative behavior in interactive work.[75] Often, these experiments challenged perceptual habits and expectations (left-hand menu bars, forward and back buttons) by creating interfaces that were purposefully difficult to use and navigate. The Japanese designer Yugo Nakamura's 1999 website, Mono*Crafts (figure 4.2), for example, employs a navigational interface system that demands a fine degree of mastery on the part of the user, who must struggle with gaining control of the speed, direction, and scale of a moving strip of interface buttons by learning to anticipate the way the interface responds to mouse movement. It is difficult for a user to place a virtual finger on a category because the interface moves so fast in the opposite direction, creating a sensation of trying to grasp something slippery. If the user moves the mouse to the left, the interface moves right; moving the mouse to the right causes the interface to move left. Moving the mouse to the top of the screen pulls the interface in closer, while moving to the bottom pushes it farther away and turns the strip into a tiny, unreadable line. The widespread circulation of personal experimental spaces such as this generated genuine excitement for the possibilities invoked by these new ways of visually representing and interacting with screen space.

Numerous collaborative spaces emerged for interactive artists to share experimental work created with Flash and Shockwave. The Remedi Project (REdesigning the MEdium through DIscovery), a gallery site for digital artists that ran from 1997 to 2002, was "created in the belief that by suspending judgment about the web, and by abandoning our preconceptions about how to use it to communicate, we may find a better way to express our discordant voices." Co-creator Josh Ulm explained, "I'd like to see people start telling stories with the Internet. With interactivity. I don't think much about video on demand and online shopping. . . . I'd like to see people collaborating and building ideas that we never anticipated or haven't even dreamt of."[76]

For many designers who worked on client projects by day and built their personal sites and experimental projects by night, Flash was more than a way to market edginess, specialized skills, and creative authority; it was also a way to intervene in the transactional discourse of the New Economy by shifting attention from shopping carts and secure servers to new creative forms that emphasized storytelling, collaboration, and experimentation.

Figure 4.2. Screenshot from Yugo Nakamura's Mono*Crafts interface (1999).

The software was easy to pick up—creating a Flash animation required no familiarity with programming or even HTML code—but the high learning curve for more advanced interactive techniques meant that mastering Flash required being immersed in a community of designers and developers. The socio-technical networks that articulated Flash to a particular way of conceiving, styling, and executing the production work of web design cohered through a number of community venues. Some of the more prominent spaces for circulating shared notions of quality and prestige were the handful of "design portals," collected and edited links to the "best" implementations of Flash, which served as regularly updated global entryways into the Flash design scene. Joe Shepter, a writer who covers the design community explains, "Every day, dozens of sites get updated, and dozens more get launched for the first time, each one with high ambitions of getting thousands of visitors and a spot at the podium of the next Web design conference. To keep your finger on the pulse of the movement, the best thing to do is find yourself a portal. . . . Their editors sift through hundreds of emails a day from designers around the world, and post the links they like to a news board."[77] Portals intensified competition among practitioners and tended to promote a culture of (mostly male) highly venerated "star" designers, whose styles were copied profusely.

The award culture of Flash has roots in the Cool Site of the Day phenomenon of the early web but was also inspired by Macromedia's practice of designating a Flash "Site of the Day" on its website, a highly regarded honor for the featured website. The explosion of film festivals and conferences dedicated to Flash helped generate shared ideas of innovation and creativity as sites were nominated for awards in categories such as motion graphics, typography, navigation, sound, video, 3D, and application development. The film festival component to these awards shows embraced ideals of the film auteur, positioning web design as the artistic expression of a director with a unique vision. One judge of the *Communications Arts Interactive Design Annual* awards in 2000 was asked, "How do you see the skill sets of designers evolving as more bandwidth becomes available?" He responded, "As the Web moves more towards broadband it will require less of a designer and more of a director who understands [the language of] 'cut' and editing and flow of information. . . . The experience can reach emotion; there's anticipation while you maybe watch a special loading screen and then there's a build up. Things like that, . . . once they're incorporated, will pull more of a viewer connection to the information."[78] This shift from "designer" to "director" suggests a number of important revisions to web practice. It places the user experience firmly in the hands of the web auteur, who strives to create an experience that can "reach emotion." The web user becomes a viewer, a spectator whose main purpose is to be enthralled, to feel anticipation in the buildup of motion graphics and introductory sequences.

The film festival atmosphere of Flash conferences helped position the designer as an auteur, but it was often the crowd that reinforced and celebrated the comparison. Many attendees have described the Flash film festivals as stimulating, exciting, and inspirational; seeing the "faces behind all these big Flash names," as one attendee put it, lent an air of celebrity and adulation.[79] But unlike in the film industry, the rise of Flash auteurs helped new designers believe that their status as future Flash superstars was entirely in reach. This was a self-made medium, and star designers were regarded as a symbol of what was possible if you master the software and use it to harness your creative freedom. Many walked away from Flash community events anxious to delve into the possibilities that they witnessed. This account of one designer's first experience at the Flash film festival in 2000 demonstrates this intensity well:

> The energy was electric as the best of the Web was projected on a large screen to the sounds of cheers from the audience. Awards were presented and designers and developers were recognized for their talents and contributions. I was blown away. It was a palpable sense of community, and it engaged me.... I now knew what I had to do. It was clear to me as I drove back to Boston with my head spinning faster than the wheels on my car. It was time for me to immerse myself in the world of Flash design and animation. I needed to push my Flash skills further and build my Flash Website.[80]

These conferences, portals, and Flash community sites also proliferated certain visual styles that had become the province of Flash auteurs: complex transitions, reactive graphics that respond in real-time to mouse movement, and cinematic opening introductory sequences that soon became associated with quality and craftsmanship.

One website that exemplified this auteur-inspired Flash aesthetic was the site for the upscale department store Barneys New York, designed by the Manhattan web studio Kioken in early 2000. With Barneys' choice to launch its first website as a Flash interface instead of an e-commerce vehicle, the Barneys site relied on motion graphics and novel forms of navigation to signify the characteristics of luxury and high fashion while maintaining a privileged distance from the masses that e-commerce served. When inventory is suddenly available twenty-four hours a day, seven days a week, to shoppers who may live far from the urban centers where upscale retail stores are located, shopping becomes more accessible and democratic, catering to all internet users, not just those who live in fashion capitals around the world. By contrast, the Barneys site kept viewers at a polite distance from the images displayed on the screen. Viewers could look and admire, but there was no shopping cart or checkout line to translate visual desire into personal ownership. One would have to visit the store for that.

As soon as the website loads, transparent blocks of color masking high-resolution cropped images of models' bodies in tailored garments

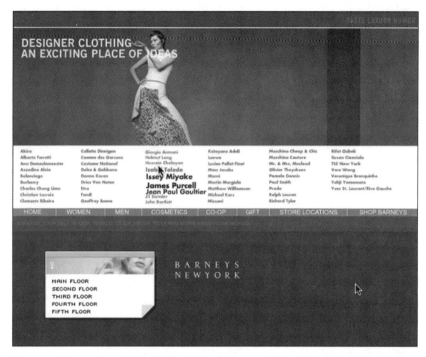

Figure 4.3. Screenshots of the Flash website for Barneys New York, designed by Kioken in early 2000.

slowly pan across the screen. When a user clicks a navigational element, the on-screen image immediately reacts by dashing off the screen. The interface pans to reveal a new screen space that slowly slides into place, styled with motion graphics and small, draggable windows that can be arranged by the user in the lower space of the screen. Each time a new category is selected, the interface instantly ushers previously off-screen images into the viewable frame, where they slowly come to a halt. This navigational device of constant panning across partially obscured images of fashionable clothing and glamorous models encouraged a scopophilic mode of looking in which the viewer has a semblance of control but the images are still out of reach. The interface positions the viewer as a spectator at a runway show; the web page never leaves, never reloads—clicking just moves the field of vision by sliding frames into view from offstage. Vertical translucent blinds become a stylistic device that gives these transitions a showman quality: fashion appears from behind the curtain and makes its way across the stage. Partially opaque text at the top of the screen reads, "Taste Luxury Humor," hailing users to read these images as a high-quality performance, one that is cheeky, fun, unexpected, and above all, tasteful. The combination of motion graphics and draggable windows merges a cinematic sensibility with human-computer interaction to simultaneously highlight a cinematic distance between the spectator and the image while also introducing a newfound freedom to grab an object on a website and move it across the screen. It is a gesture that moves the logic of the web from the page to a hybrid space between film, theater, and desktop metaphors.

The new formal qualities of the web introduced in Flash sites such as Barneys' became widely circulated, quickly emulated, and highly valued by inspired protégés enamored with star designers and the art of Flash design. As the commentator Jeffrey Zeldman wryly notes, "The Web used to look like a phone book. Now much of it looks like a design portfolio. In fact, it looks like the design portfolio of 20 well-known designers whose style gets copied again and again by young designers who consider themselves disciples."[81] This widespread circulation of Flash design and the notions of quality and value attached to particular expressions of the Flash aesthetic helped to give shape to another "New Economy"—a "New Visual Economy" that linked talent, social prestige, and high earning power through the circulation of Flash images, generating a mode of value that challenged the web of e-commerce. By 2000, accomplished Flash designers could earn $2,000 a day for Flash-related work and top auteurs could often earn significantly more than that.[82] For those who were invested in content rather than transactional websites, Flash represented a return to a creative core that many felt had been lost with the turn to e-commerce. Although plenty of e-commerce sites incorporated Flash content, and plenty of Flash sites incorporated transactional processes, these two visions of the web at the height of euphoria represented a clash of logics between the New Economy of transactions and the New Visual Economy.

The transactional web, powered by the uninterrupted flow of constantly updated, database-enabled, real-time inventory lists and purchase-processing capabilities, was losing its connection to the process of artistic creation. Transactional sites were functional but utilitarian; the aura of something artistic or unique, a hallmark characteristic of the early web, was fading. As Miège has pointed out, this link to the creative process is imperative to the production of audiovisual media content.[83] As the web of content pushed up against the web of transactions, the artistic process needed to be renewed. The small, specialized boutiques that were developing edgy ways to interact with visual space were doing this by rewriting the rules of design, presenting the New Visual Economy as the future of the web, one that big conglomerates and e-solutions consultancies just did not get. The emergence of the Flash auteur, then, was partly a reflection of the fandom surrounding the software and the new creative modes of use it engendered, but it was just as important to clients, agents, and interactive development firms desperate to tap into the web's new visual culture and all it stood for. Commercial clients hoped to attach a creative edge to their online initiatives and hired talented Flash designers who could visually translate brand interaction into a sophisticated and engaging experience.

Perhaps the most iconic Flash auteur was Joshua Davis, an interactive artist who worked as Kioken's senior design technologist from 1998 to 2001. Davis became famous for the work he developed in his personal websites PrayStation and Once-Upon-A-Forest (figure 4.4), where he experimented with sound, color, generative art, and navigational fields. These experiments often found their way into commercial projects. The navigational system for Barneys, for example, was hashed out through PrayStation experiments and offered to the design community as downloadable source files complete with instructions detailing how Davis was able to achieve the effect of an image slowly panning into place. Designers hungry to learn more about Flash quickly gobbled up these files and the techniques they demonstrated. Davis was the headline speaker at numerous Flash conferences, his personal sites were repeat winners of Flash film festival awards, and he was offered contracts for books and book chapters in which he detailed his creative and technical processes. In *New Masters of Flash* (2000), he further deconstructed the navigational system he created for Barneys and described how his views on chaos theory fit into his Flash development process.[84] He marketed and sold CD-ROMs of his hard drive, presented in a custom plastic case designed to resemble a PlayStation game console, which contained nearly 4,000 "original source files, art files, text files, accidents, epiphanies, etc."[85]

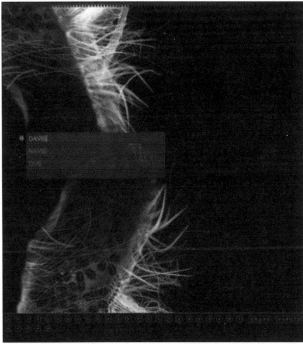

Figure 4.4. Examples of Joshua Davis's work for his personal site, Once-Upon-A-Forest (2000)

The popularity of Davis's personal sites and the huge attention his experiments generated inspired him to rethink some of the traditional rules of web design, which typically focused on intuitive navigational systems and architectures that prioritized the orientation of users in a familiar and comfortable information space. In a July 2000 CNN story that featured Kioken in a segment titled "Window on the New Economy," Davis explains,

> Most people in the commercial world will tell you that you have approximately two seconds to capture [the audience's] attention, so a site needs to be designed with a certain intelligence of navigation, it has to be easy to use, the information has to be accessible. Because you have *two seconds* to capture their attention, and if you don't, they're gone. So I had designed this site that was the exact opposite of commercialism; it was hard to find, it's hard to get to, there's no help, no instructions. And the average user spends thirty minutes on my site, and after the user is done, they mail me wanting to know if they *missed* anything.

Applying these lessons to corporate projects such as the Barneys site, Davis's work emphasized exploration over usability. It was an approach that many corporate clients eagerly embraced: the "opposite of commercialism" was desirable because it represented a way to engage web-savvy, media-conscious audiences with interactive experiences that were sticky, edgy, and unique. By pushing the boundaries of a user's expectations, Flash could help assure sophisticated audiences that these companies, and the brands they promoted, did in fact "get" new media. As the visionaries powering the future of an innovative, envelope-pushing web practice, Flash auteurs were able to assert creative power at the height of euphoria, challenging big media conglomerates and the massive e-consultancies in the process.

The CNN story frames the designers as a "band of code warriors" who are "rewriting the rules of design" and "rewriting the rules of business in the process." The websites profiled, for Sean "Puff Daddy" Combs and Jennifer Lopez, show a cinematic web in motion, set to background music and laced with hypervisual transitions and typographic effects. "We're not trying to make the web of today. We want to show what is possible," explains Kioken co-founder Peter Kang. "We want to build the future of broadband," adds Gene Na, the other co-founder. In this way,

Flash sites represent a new way of imagining the future; it is a future unconstrained by download time and focused on content—"high quality," "cutting edge," multimedia content. This future-oriented approach takes Macromedia's trademark slogan—"What the Web Could Be"—and gives it a very specific cultural form. The web could be like television or music videos or film—but with a new interactive twist. It did not have to be like brochures, magazines, or online shopping catalogs.

The power that small studios such as Kioken were able to command by wielding Flash in the era of dot-com euphoria was not just a metaphorical power that celebrated the freedom to create. The demand for top creative talent helped support an industrial environment where visual design and innovative forms of creative expression were so highly desired that small boutiques were able to handpick their clients and even go head-to-head with massive global corporations whose web strategies they disagreed with. Kioken gained a reputation in the spring of 2000 as the quintessential new breed of web designers whose "artistic whims," as the CNN feature describes them, take precedence over the satisfaction of corporate clients. The attention stemmed from a May 2000 article in *Industry Standard* that explored the emergence of "small, agile companies run by people who are determined to change the very look of the World Wide Web."[86] The "rebellious" designers featured were exploiting the "money is no object" internet economy for their own sake, creating "groundbreaking," graphically innovative corporate sites on terms of their own making. The popular anecdote of how Kioken fired its client Sony, because Sony was not listening to Kioken, demonstrates the extent of the power and hubris that surrounded Flash auteurs at the moment of the bubble's collapse. Sony representatives were very "rude with the design studio," Na told the astounded CNN reporter, Charles Molineaux, in "Window on the New Economy." "They ask us to do terrible work and make decisions that we thought were really not the best for them." In an environment where clients are a dime a dozen but highly skilled creative talent is hard to retain, the business model of creative web production shifts to the workforce. "That's why we have the business model we do," Kang explains. "The focus is on the employee and really catering to their needs. If we lose our employee workforce, we don't exist as a brand."[87]

When Joshua Davis was asked about client relations at the Flash Forward conference that summer, he relayed the story of the CNN interview and the studio's willingness to fire clients such as Sony and the rapper Puff Daddy, who hired Kioken to produce the website for his Badboy record label. "The problem is that you have companies who come into a studio, and they have acquired a skill set of buzzwords that they feel they need to say in a meeting because they think they know what they're talking about. And we basically said, 'The less you talk, the better,'" he said to a huge round of applause. "And CNN came because they were shocked we would turn away a big company, where traditional business, you would put up with the bullshit." In the end, Davis conceded, both Sony and Puff Daddy came back and bowed to the studio's creative vision. The excitement in this moment of transposed power was not lost on the Flash Forward audience, which cheered wildly at the news that creativity had trumped corporate bullshit. "We've been working in this medium for a long time," Davis concluded. "They might be great at building electronics or rapping, but we build websites."[88]

Indeed, at this point, there was nothing traditional about web design in the New Economy. The audacity of firing big-name clients was mind-boggling to commentators who latched onto the point that overconfident "hot shot" designers were ruining the web with proprietary technology aimed at elite viewers with broadband connections. But the conditions surrounding creative design work during this period of euphoria also demonstrate something quite uncanny. Media moguls have addressed the oversaturation of media and entertainment alternatives by using key strategies such as branding to develop a collection of products that are linked across media.[89] Yet here was Sony, a major media company, unsure of its footing, deferring its vision to a new breed of designer who calls the shots and charges $100,000 a month just for Sony to remain a client. In this environment, edginess and creative vision became articulated to the "future of broadband," what the web could be. The discourse of the New Visual Economy served as a powerful rebuttal that small boutiques harnessed to win back some of the power that had been ceded to global e-commerce consultancies pursuing a transactional vision of the web.

Conclusion

By exploring the discursive formation of the New Economy during the euphoria period of "internet mania," this chapter has highlighted how different visions of the future of the web played out against a backdrop of venture capital, stock-market gains, industry consolidation, and competing modes of web practice. Cultural industries, Miège suggests, occupy a zone of continuing struggle and interweaving logics as actors seek to protect their interests and put themselves in a strong position for the emergence of new media. As he describes it, clashes between the flow model of broadcast and the editorial model of books, records, and film play out in a number of ways: between firms and investors, between artists and other members of artistic professions, and in how products are conceived in production processes and in modes of consumption.[90] While e-commerce and Flash are both web-enabled practices, the different ways of conceiving, financing, producing, and imagining the web likewise represent competing logics that interweave and clash. The tensions between these two discourses of the New Economy— one constructed through transactions, the other through visual experiences—reveal fundamentally different ways of conceiving value, expertise, quality, and the user. I have argued that Flash web design in this moment of euphoria was both an intervention in the New Economy discourse that privileged the web as a space of commercial transactions and a rearticulation of that discourse to a New Economy of images, a New Visual Economy based on the circulation of a privileged Flash aesthetic. Both versions of the New Economy, as a business model and a visual system, depended on the mantra "the old rules no longer apply."

The wave of conglomeration that accompanied the pursuit of the "one-stop shop" and the emergence of the web auteur as a way to imbue cultural artifacts with a renewed sense of creativity were both linked to the power struggles and industrial practices that accompanied web euphoria. However, these processes are not without historical precedent. The rise of the film auteur in the late 1960s, it should be noted, was bound up with industry conglomeration, technical opportunities, and marketing strategies in a moment when the industry was facing crisis and revision.[91] Likewise, the rise of the producer-auteur of network television came to prominence under the threat of cable and the need to align broadcast fare with "quality" programming.[92] Web auteurs responded to similar industrial and creative pressures, employing new modes of production and appeals to authorship to cement notions of quality in cinematic terms that challenged the dominant discourse of e-commerce.

5

Users, Usability, and User Experience (2000–2005)

If you could go back in time and unload your shares of dot-com stocks at the very height of the market, then Friday, March 10, 2000, would be the perfect time to sell. The NASDAQ closed above 5,000 for the second day in a row—nearly double its value from a mere five months earlier. It would not reach that peak again for another fifteen years. For the next month, the stock market was incredibly volatile—dipping, then creeping up again, only to take another drastic plunge days later. By the end of March, the NASDAQ had lost 11.7 percent of its value, surpassing the 10 percent benchmark Wall Street uses to define a stock-market "correction."[1] April was much worse. Although accumulated losses topped 20 percent (at which point Wall Street designates a bear market), many stunned and shaken investors were still not sure the bubble had actually burst. "It's a long overdue re-evaluation in the new-economy stocks," said the chief investment strategist at A. G. Edwards in St. Louis. It is not a bubble bursting, he contended, but a fairly normal downturn that "reduces excesses and complacency" and should prove generally beneficial.[2] When trading closed on April 14, 2000, after a day of relentless selling, investors had lost more than $2 trillion in paper wealth in a single week, and the market was down 35 percent from its March 10 peak.[3] In hindsight, the numbers revealed that the bubble had in fact burst. But as John Cassidy notes, for almost a year and a half after the NASDAQ's April crash, there was a marked reluctance to accept that the long boom had come to an end.[4] After all, the market had fallen—drastically fallen—in the past but always bounced back. The losers were those who cashed out; the winners "buy on the dips." As the *Wall Street Journal* pointed out, despite such heavy losses, the NASDAQ was *still* up 32 percent from the year before—hence, "the declines pale against the gains they have racked up in recent years."[5]

Commentators pointed to a number of possible triggers for the mass sell-off, but it was still hard to pin down exactly why the panic began when it did.⁶ Much of the blame for the hype fell on the mainstream-media cheerleaders and bullish investment analysts. But the alarming sell-off was largely attributed to susceptible, impulsive, novice investors. An article in the *New York Times* observed, "Main Street began to drive Wall Street, inverting the power structure and dethroning brokers, investment bankers and money managers as ultimate market arbiters. . . . Welded together by Internet chat rooms and nonstop cable television coverage, legions of retail investors have formed a critical mass of opinion, rudely casting aside the wisdom of such value investors as Warren Buffett. The widespread involvement of ordinary Americans has made this bubble more volatile and hazardous than its predecessors."⁷

Wall Street's "dethroning" by ordinary Americans might be the ultimate expression of that celebrated New Economy axiom "the old rules no longer apply." Out with the experts! Wall Street, however, was not about to relinquish its claim to knowledge or financial expertise so easily; instead, financial analysts took it upon themselves to "temper the wild excesses of new investors."⁸ Mostly they did so by sounding the call for "a return to fundamentals," to considerations of assets, cash flow, returns on equity, and dividend yields as indicators of value.⁹ Companies such as Cisco and Oracle have "fundamentals," while other companies, "the sexy, flashy ones with the right buzzwords," do not.¹⁰ Smart investing involves sorting the wheat from the chaff. From this perspective, at least in April and May 2000, the correction was framed as a positive development, a long-overdue adjustment that would restore some health to the market's long-term prospects.¹¹ This was not the end but a new beginning, one in which sound business principles would hold sway over the vacuous hotshot startups of the recent past. By summer, the NASDAQ was back to over 4,000, massaging hopes that a recovery was well on its way.

The optimism, however, did not last. The summer gains were eliminated by September, and stocks steadily declined through the fall of 2000. Once-prominent dot-coms were bankrupt or trading for pennies. Pets.com announced it was closing shop that November, just seven months after its IPO. TheGlobe.com, which made a huge splash in November 1998 with an IPO that opened at $9 and closed at $63.50 on its first day of trading, was trading around $0.63 a share just two years later.¹² Larger e-commerce players such as Amazon announced major job cuts and closed plants, while the web integrators and e-consultancies that advised dot-coms and built their websites struggled to stay afloat.[13] The day before the terrorist attacks on September 11, 2001, the NASDAQ closed at 1,695.38, and the index dropped another 200 points the week after the market reopened. A year later, in autumn 2002, the NASDAQ dropped below 1,200. That was when the market hit the bottom, and by this time, it was very clear that the internet boom was long over. But it was not just paper wealth that had vanished. Alongside the bankruptcies, million-dollar websites disappeared into the digital ether. Domain names were vacated and resold, leaving little evidence of earlier-generation websites. Companies once eager to play up their association with the internet were seeking distance instead. Hundreds of thousands of layoffs followed, leaving web workers to mingle at "pink slip parties," networking events designed to help laid-off workers find leads in the more frugal, internet-wary climate of the post-dot-com job market.[14]

How did this volatile economic context affect the labor and cultural meanings of web design? This chapter analyzes how the material semiotics of the web—an ensemble of production practices and technologies along with their entwined imaginative dimensions and symbolic expression—were reconfigured in a sober, postcrash environment from an assemblage that I have been calling *dot-com design* into a new sociotechnical arrangement that cohered around a design paradigm known as *User Experience* (UX). If the market had become wary of unproven dot-com startups that promised to transform the world through online groceries and pet-food delivery services, a parallel wariness accompanied the profusion of lavish websites designed with Flash, the animation software program web designers were using to create full-screen motion graphics, interactive games, and rich media advertising. How different the mood was from the dot-com euphoria of 1999! In chapter 4, I mapped some of the industrial shifts taking place at the very height of the bubble (from late 1998 until early 2000), when the internet was abuzz with news of endless mergers and acquisitions that fueled a frantic race to assemble global, full-service, one-stop-shop mega-agencies. Within that high-stakes environment, I argued, there was a tension between two different visions of the web. One was a web of transactions, seen in the sudden rise of online retailers invested in electronic commerce and the e-business consultancies that emerged to help steer these new internet-focused businesses through the transformations associated with the New Economy. Another vision was expressed through the proliferation of Flash content online, which privileged a New Visual Economy of creative expression and sensory experience over the transactional discourse of e-commerce. The conflicts between these visions were not merely the result of particular business strategies or aesthetic concerns but instead represent sites of struggle over the very meaning of the web—what and whom it is for, how it ought to be designed, and which experts are best equipped to advise on such matters.

Picking up where that chapter leaves off, here I follow the discourse surrounding Flash through the bursting of the internet bubble and its aftermath. As the notion of the web auteur came under fire for stylistic excess, a new dominant discourse of "usability" gained traction, a discourse that downplayed the designer and emphasized the user. In this moment, quality web design was less a cinematic experience for viewers or an experimental venue for designers; it was easy, efficient, and above all, usable. Whether tempering the wild excesses of the market or the interface, the "return to fundamentals" that followed the bust involved disciplining overindulgence and hysteria with new sets of rules for evaluating worth. In the world of stock-market speculation and the world of web design alike, emotion and affect were to be toned down in favor of rationality, science, and user testing. This heightened concern with the practical matters of function and efficiency, however, later earned criticism for course-correcting too far by devaluing the aesthetic experience of interactive environments entirely. What had become of the "feel" in designing the "look and feel" of a website? Could perceptions of a website's usability be impacted by aesthetics, affect, or emotion?

Between 2001 and 2005, web design best-practices discourse began framing usability as just one component of the big picture, and that was "User Experience." The term was expanded from its more narrow application in the field of human-computer interaction (HCI) to a much more holistic application that positioned UX as integral to product management and customer service. Within a decade, corporations came to see each and every interaction a customer has with a product or service (called "touchpoints") within the purview of UX. This could occur online or offline—from the passing highway billboard to the in-store display to the customer-service phone tree to the unboxing experience to the Twitter account to the design of the website—the entire fabric of *experience* involved in interacting with a product or service does not stop or start on the internet. Today, UX is no longer just an issue of information architecture or even interaction design; it is increasingly infused within the organizational logic of twenty-first-century enterprise, where it is seen as integral to ensuring lasting relationships between customers and brands.

In the first few years immediately following the dot-com crash, the discourse of UX was harnessed by different stakeholders and used to help reconcile the competing values associated with usability on the one hand and rich media experiences on the other. In doing so, I argue, it forms the emergent care structure of the postcrash economy, in which the internet and the web are no longer the province of geeks and early adopters but increasingly integrated into the everyday lives of a more general public. If dot-com design was an expression of the spectacular and the extravagant, User Experience design gestured toward the ubiquitous, ordinary experiences that make everyday life more manageable. To follow this trajectory, I begin by turning to the redesign of the Flash MX authoring software in 2001, analyzing how the Macromedia development team internalized the critiques leveled in the wake of the dot-com crash and relaunched a product that, by design, would assist Flash producers in the transition from inspired amateur spinning animated intro sequences to seasoned technologist engineering "architectures of participation." The reconfiguration of Flash from a web animation software package to a development platform for managing "rich internet applications" (RIAs) demonstrates how swiftly (though not always smoothly) software applications are repositioned in response to economic and industrial pressures, even as the socio-technical imaginaries surrounding them are less easy to control.

Turning from the redesign of software to the reorganization of the web industry in the wake of the internet shakeout, the second case study examines how the tensions between usability and experience design resulted in a new breed of specialist, the UX consultant. I focus on the formation and growth of Adaptive Path, a user-experience consulting firm co-founded by seven dot-com veterans in 2001. Tracing the development and strategic positioning of Adaptive Path, I examine how UX came to signal "quality" web design and how, in the process, both the meaning of the web and the role of design were articulated to a new cultural imaginary that promoted ordinary users over star designers, participation over publishing, and sharing over surfing. This ascendancy of the user was part of a broader discourse that connected material practices and modes of production (how a website should be planned and functionally implemented) to broader ideologies and cultural meanings (how participation is democratic, how amateurs are empowered) that came together as a potent new vision of the web: Web 2.0, the user-driven web. While a number of scholars have drawn attention to the ideological dimensions of Web 2.0—as a mechanism of status and microcelebrity, as a neoliberal marketing logic, as a form of alienated labor, and so on—this chapter historically situates the rise of Web 2.0 (2004–5) in conversation with web-building communities advocating for web standards and user-centered design methodologies.

The Flash Usability Campaign

To understand how Flash was challenged and reimagined after the crash of the dot-com bubble, it is helpful first to turn to some of the literature that explores the way social interactions surround and influence technology design and technological change. Constructivist accounts typically emphasize the development stage by focusing largely on how new technologies arise, are negotiated, and then are finally stabilized. Feminist and media/cultural studies scholars have pointed out that limiting concerns to the conception, invention, design, and development of a technology forecloses a more nuanced engagement with the ongoing negotiations surrounding meaning, power, and ideology. For example, Hughie Mackay and Gareth Gillespie suggest analyzing technology not solely as design processes but also as functional and symbolic systems that are informed by ideology and encompass both encoding and decoding. They expand the idea of "social shaping" toward a matrix of practices that include the marketing and advertising of technologies and their social appropriation by users who reject, redefine, customize, and invest their own meanings in them.[15] Ron Eglash identifies a number of ways in which people outside the center of social power appropriate technologies by changing their usages, meanings, or structures for their own kinds of technological production. This may involve changing the semantic association of a technology, adapting a technology by violating its intended usage, or even reinventing a technology by creating new functions through structural change.[16]

Technology (and software in particular, since it is programmable) offers a certain "interpretive flexibility" through the wide variety of designs and uses that are possible.[17] To interrogate the possibilities and potential for different ways of using and making sense of technology, Steve Woolgar suggests a "technology as text" approach that addresses the interplay between social dimensions and designed cultural artifacts.[18] As a text, technology can be understood as something manufactured, designed, and produced within a particular social and organizational context. As such, it likely embodies specific goals and values while leaving open an "irremediable ambiguity" about what the technology is and what it can do. But, Woolgar explains, this is overlaid by sets of preferences that guide the "appropriate" interpretation and use of technology. Former users of Microsoft Office 2000, for example, may remember the Office Assistant Clippy, an animated paperclip that periodically popped up to inquire if the user was writing a letter and needed help. Thankfully, Clippy was easily dismissed ("Don't show me this tip again"), and users of Word software could compose all sorts of documents besides letters. Yet the interface icons, help files, templates, metaphors, and marketing all orient users toward an application in particular ways. Creating a new file from one of Microsoft's preformatted templates offers a gallery of options: business cards, fax coversheets, real estate newsletters, school calendars, garage sale signs, and bistro menus. These templates are not just library items with predefined fonts, margins, and layout styles; they embody scripts that delegate roles, responsibilities, and actions to users and technological artifacts.[19] By virtue of how users are addressed, they are interpellated into certain subject positions. In a critique of user-friendliness in corporate network environments, Alan Liu notes the plethora of Office template options and quips, "Where, by contrast, is the icon or template for 'Whistle-Blower,' 'Sexual Harassment Complaint,' or 'Discrimination Grievance,' let alone 'Unemployment Benefits,' or 'Welfare Application'?"[20] These alternative formats fall so far afield of the user-friendly scripts of word processing that Liu's question is funny because it is so preposterous. But it also reminds us that preferred meanings are not just embedded in software; they can be supported (or challenged) through intermediaries—marketers, corporate interests, consultants, journalists, academics, artists—who occupy strategic positions between the producer and user of information technologies and may therefore intervene in the interpretation or reading of a text.[21]

So although reading against the grain is always possible to some extent, it is often kept in check by powerful incentives that invite audiences to inhabit anticipated roles that sanction particular uses of a technology while subtly discouraging other practices. As Woolgar notes, "ways of using software other than those the designers had in mind are possible, but they turn out to be prohibitively costly (since alternative sets of material resources will be needed to counter or offset the effects of the technology) and/or heavily socially sanctioned."[22] Sometimes preferred readings can be rejected; at other times, technologies fix or lock particular paths into place by hard-coding certain ways of engaging with software or forbidding other uses. As Tarleton Gillespie has argued, the design of a software product and the powerful narratives surrounding it choreograph the actions of its use, building in roles for users to play and paths to follow, subtly rewarding users for inhabiting certain subject positions. Attending to technologies' "affordances," Gillespie reminds us, involves paying attention to the assumptions a given tool makes about the user and the types of activities that are encouraged and endorsed, as well as those potential uses that are physically shut down or prohibited by the actual design of the technology.[23] As we will see with the broad campaign by Macromedia and other key participants to change the way designers used and imagined Flash software (and, by extension, the web itself), the urging of a preferred reading was by no means subtle. The loud, pleading cries for change are a useful reminder of how hard it can sometimes be for industries and organizations to intervene in the symbolic practices of users.

"Flash: 99% Bad"

For Flash design, the summer of 2000 was what some designers later recalled fondly as the golden age of Flash.[24] As one Flash designer reminisced in a blog post,

> there was a period in early [19]99 when todd purgusen (juxt) and josh ulm (remedi quoka) were gods and dot com dollars were everywhere . . .

thats when I started using flash . . . exciting times. then flash 4 came out and by early 2000 the first flash forward happened and flash had really arrived . . . brenan dawes psycho studio . . . the praystation experiments . . . yugo . . . the flash kit arcade . . . amazing. I can recall staying up very late (Im in Australia) reading posts about the flash 5 launch at flash forward in ny [in July 2000], it seemed like the greatest time to be working with flash.[25]

As this poster and countless others have noted, Flash masters were revered like "gods," and the demand for tutorials, open source files, interviews with Flash masters, and inspirational Flash sites reached an all-time high. In fact, if book sales are a useful indicator of interest in a particular technology, then if graphed accordingly, Flash mania in web production circles might even be mistaken for the NASDAQ, with a similar explosion of growth between 1999 and 2000. Flash titles made a significant showing in Amazon's "Bestselling Computer and Internet Books of 2000" list. The book buyer for the bookstore chain Borders remarked in early 2001 that "Flash was THE hottest topic last year and still is this year."[26] The UK web design magazine *Cre@te Online*, which launched in July 2000, regularly showcased Flash sites in stories that "spotlight the very best sites on the Web today." Flash was featured in design perspective columns such as "The Software I Couldn't Live Without."[27] In fact, of the ninety-six websites showcased in the first six issues of *Cre@te* (July–December 2000), eighty-five were created entirely in Flash or prominently featured Flash content.[28] For creative designers, Flash had become the de facto standard for building high-quality and cutting-edge web-based experiences.

Yet it was not too long after the stock market reached its tipping point that the discourse surrounding Flash began to register a noticeable chill. In October 2000, the usability "guru" Jakob Nielsen's influential critique of Flash attracted significant attention. In his unambiguously titled article "Flash: 99% Bad," Nielsen lambasted the software for "encouraging design abuse," "breaking with Web fundamentals," and "detracting from a site's core values"—charges expressing a biting exasperation with the bells, whistles, and lengthy animated intro sequences that were seen as part of the "look what I can do" bravado of the Flash aesthetic. This aesthetic, Nielsen noted, was not one founded on the interactive capabilities of hypertext. "Unfortunately, many Flash designers decrease the granularity of user control and revert to presentation styles that resemble television rather than interactive media," he complained. In his view, Flash was mere "fluff" created by art-school graduates with overinflated egos. "You still see a lot of silly Flash intros, for example, which are just there to make the designer or website owner feel good. But we all know users just click the 'Skip Intro' button right away because they just want to get to the facts."[29] Nielsen argued that users "have a utilitarian attitude" toward most of the sites they visit; because they are there to seek information, they "just find the graphics distracting."[30] To best meet users needs, Nielsen offered a number of "laws" that every website should follow so users could quickly and easily find what they came for. One key to accomplishing this, Nielsen insisted, was that "Websites must *tone down their individual appearance and distinct design* in all ways: visual design; terminology and labeling; interaction design and workflow; information architecture."[31] Each site on the web, he believed, should have a consistent look and feel; this was the only way users would be able to intuitively and quickly move across the vast terrain of the web. In his July 2000 article "The End of Web Design," Nielsen admonished designers for wanting to create unique experiences; he urged them to "relinquish control" in the service of creating the web as a "unified" whole. The very first law, "Jakob's Law of Internet User Experience," states, "Users spend most of their time on *other* websites." Therefore, he reasons, "users prefer your site to work the same way as all the other sites they already know." Designers, he said, should "think Yahoo and Amazon. Think 'shopping cart' and the silly little icon. Think blue text links." If these symbolic systems were always consistent, the web would be more reliable and more usable.

Designers were outraged. They countered that such a formulaic "template design" would strip the web of the character and expression that creative individuals contribute by crafting distinctive experiences. As the noted design instructor Lynda Weinman put it, "I think there is a real danger in being too formulaic because then it stops being interesting. If you look at the work of famous artists or writers, anyone who has made their mark has done something different to stand apart from the rest."[32] But for Nielsen, web designers were not traditional artists who created "final" works like paintings, sculptures, or films. Rather, users created the final work by ultimately "composing their own experiences" through personal mouse clicks and pathways around the web.[33] By placing the user above the designer, Nielsen dealt a hard blow to the idea of a web auteur. Joshua Davis challenged back that Nielsen's ideas not only stifle growth and creativity but portray the user as a dupe. "He wants to hand hold everyone through their experience of the Internet," Davis complained. "It's like saying everybody's stupid and that if the average idiot can use a website, then a professor can use it too. But I don't want to assume that everyone is the average idiot."[34]

Like other battles between competing visions of the web described throughout this book, the struggle between Flash designers and usability proponents was partially a contest over the value of professional skills and perceptions of expertise. But alongside a plummeting and unpredictable stock market, anxiety and passion collided with vehemence; countless rants about the sad state of things and who was to blame dominated the email lists, discussion forums, and trade magazines devoted to web design and development. The impact of the "Flash: 99% Bad" article can hardly be underestimated—even a decade later, it was still being widely cited in web design circles as a monumental turning point in the discourse surrounding Flash. But for all of the notoriety that the article attracted, it was not the first time Nielsen publicly criticized Flash as a "gimmick" that produced "gratuitous animation."[35] Nor was Nielsen the first to declare that "Flash is evil"[36] or that it "constituted a cancer on the Web."[37] Indeed, the once-fashionable Flash "introductory sequences" that Nielsen railed against had already been deemed passé. As early as 1998, the web designer Yacco Vijn created a popular parody of the "skip intro style," named for the obligatory button added to sites should users want to skip the introductory animation (figure 5.1). In it, viewers are forced to watch an endless array of orange balls while the site loads; underneath a message explains, "this may take forever, but hey it's an intro!" Once loaded, the site reveals full-screen rotating spheres and welcome messages, accompanied by a score of "over-the-top science fiction sound samples . . . [reminiscent of] the Gabocorp.com envy that was all the rage at the time."[38] While the site was a popular amusement during the dot-com bubble, it circulated with viral ferocity after the bust.

The awakening of such powerful anti-Flash sentiments in late 2000 coincided, of course, with the downturn of the internet economy and

Figure 5.1. Screenshot from Yacco Vign's Flash parody, "Skip Intro" (1998).

the failure of so many dot-coms to prove viable business models. But beyond a waning tolerance for the perceived self-aggrandizing of the web's Flash-based visual economy, the discourse of creative expression, exploration, and experiential design was hard to reconcile with the "return to fundamentals" that internet consultants and investment strategists proclaimed would signal the maturing of the internet economy. "Usability is being talked about a lot more by clients and the agencies," noted Mike Bloxham, founder of the digital media research and consultancy firm Net Poll, in early 2001. "Investors and companies are demanding accountability and they expect usability to help them find that."[39]

Websites that were hailed a year earlier for their edgy, sexy, flashy, ultracool interfaces were more frequently being accused of "over-design."[40] Techniques initially developed to frame the web experience in cinematic terms—such as launching a new window that took over the entire visual field of the desktop and eliminating standard browser buttons ("back," "forward," "home")—were now beginning to signify something other than cutting edge. Indeed, Flash sites were being described as the visual manifestation of irrational exuberance. The discourse of usability, therefore, provided a clear and convincing set of fundamentals that could temper the "wild excesses" of the interface. The standard blue, underlined hypertext links that Nielsen advocated were the interface equivalent of dividend yields and cost analysis—boring but necessary considerations that would stabilize the web's semiotic system. Needless to say, creative designers were not anxious to embrace this web practice. "He wants to make the Web so boring!" designers exclaimed. It was the equivalent of high-def TV reverting to black and white, others noted incredulously.[41] Besides, breaking with fundamentals was the very point of Flash, as one designer noted in an email discussion group for Flash users: "TV has honed itself over generations by breaking its 'fundamentals' and that has made it the most powerful entertainment and information force in the world. This is what we all want. We want people to watch our creations, be entertained, educated and [we want to] make a living in the process. We can't succeed in this without entertaining people, keeping them transfixed not just interested and catered to. I think Jakob Nielsen is stuck in 1995."[42] As soon as the news of the attack on Flash reached the bulletin boards and community sites, Nielsen became the public enemy of creativity within the Flash community. Joshua Davis joked in a BBC interview that he would beat him up if he saw him.[43] And only three days after the article appeared, all of the "Fuck Jakob Nielsen" T-shirts were already sold out.[44]

But the usability critique of Flash was snowballing. Accessibility advocates, long frustrated with a lack of standards that would help users with disabilities access web content, were demanding that Flash designers create content that could be accessed with assistive technologies such as screen readers and magnifiers. "If Barneys department store was to remove its handicapped ramps and bathroom grab handles, it would be sued," argued one critic. "But its inaccessible Website wins awards."[45] Likewise, web standards advocates, while still making allowances for web art and venues for creative expression, were bemoaning not just the proprietary nature of the Flash plug-in but also the growing saturation of an overdone "pixellated, techno-style" Flash aesthetic that was increasingly positioned against standards and usability concerns. Reflecting on web design practices in 2000, contributors to the web-standards-focused online magazine *A List Apart* sardonically noted, "It was a year of trendiness, inspiration, imitation, and the imitation of inspired trendiness."[46] Lamenting the "shallowness" of web design, Brandon Oelling, Michael Krisher, and Ryan Holsten observed,

> The phrase "Shockwave Flash" became the designer's poster child in Y2K. Anyone who called him/herself a designer seemed to be touting Flash's wonderful solutions to clients ("How about a Flash intro! That would be cool!!"). For web designers, Flash is a household name, but at least one study has found that most uses of Flash on the web do not consider the three most important aspects of an interactive experience: graphic design, content, and usability. Combine this misuse with an obsession to fit into the trends of design, and we have ourselves a cornucopia of websites caught at the fork in the road. The nearest street sign reads: "This way to aesthetic trendiness," and "That way to a usable interface."

The article urged designers and web agencies to recognize that "Flash is only part of the toolkit and not the only tool required to build a successful website." Ultimately, they argued, technologies such as Flash (or DHTML or JavaScript) were "merely tools."[47] By obeying the fundamentals—concepts, solid ideas—truly trendsetting projects could break through the mediocrity and drive new web experiences.

These arguments divided the Flash community. The virulent Flash backlash that took hold in late 2000 and escalated through 2001 and 2002 ignited such heated debates in web design forums and Flash community sites that moderators regularly closed down discussion threads in an attempt to restore order.[48] On the one hand, "standards," "accessibility," and "usability" interfered with the creative process. Even "masters" such as Joshua Davis acknowledged in his book *Flash to the Core: An Interactive Sketchbook* (2002) that he agreed with web standards, "but only up to a certain point." He suggested, "On the Web, we should blur the technical assumptions just that little more if it means the result will be something amazing. Perhaps the target audience at first will be only a very small, very select group of viewers. But we and they will help evolve and change the medium . . . by building intuitive systems that educate the user—not design down to the level we think the users can handle."[49] Yet, despite the creative potential for something amazing, impassioned accusations of "rip-offs," imitations, and stealing provoked bitter arguments and incited flame wars that destroyed online communities such as Dreamless, a forum that Davis set up for designers.[50] More and more Flash content was deemed gratuitous or sensational, overly concerned with glitzy effects and showing off, rather than exhibiting any real concern for people on the other side of the screen: the clients or the users.

Taming the Amateurs

By late 2000, the Flash backlash invited a lot of finger pointing: who was to blame for such bad design, usability problems, and self-indulgent productions? Macromedia CEO Rob Burgess was among the first to set the record straight. In an interview with *ZDNet* that ran the same week as Nielsen's "Flash: 99% Bad," Burgess was asked how Macromedia was addressing all of the groaning that accompanied Flash intros. He was quick to respond: "That's not the problem of the tool. That's the problem of the designer. You know who somebody is that does that? Me. I groan a lot of times at our own site and at other sites. Flash can be used for good or evil. We're trying to enunciate for the industry best practices and a lot of lessons we've learned. The laws of survival also help with that as well. Really well-designed Websites will win."[51] In this version of the "guns don't kill people, people kill people" argument, Burgess is just another web user who is as frustrated with poor implementation of this otherwise-neutral tool as the average user is. Flash is presented as an artifact with no politics.[52] "Saying the technology is bad because of how people are implementing it? That's just wrong," said the senior product manager of Flash. "You don't say that Sony created a bad television just because you have to watch programs like *Jackass* on TV."[53] Of course, televisions are used to display content, not create it. The development of authoring software is different in significant ways. For one thing, when designers interact with Flash, they use it to produce content for other users, which adds another layer of context and values between the technology and the viewer of Flash content. But how was Flash presented and packaged to speak to web designers in ways that articulated a preferred interpretation of how the tools should be used?

The Flash designer and usability advocate Chris MacGregor took offense to Burgess's accusation that placed the blame firmly on the shoulders of designers. On Flazoom, his site for Flash enthusiasts, MacGregor responded with the article "Attention Macromedia: I Will Not Be Your Scapegoat." His furious retort pointed the blame firmly back to the company that made the technology. "Macromedia has hardly lifted a finger to stop the flood of poor Flash solutions on the web," he wrote. In fact, he argued, they have encouraged bad design by awarding Site of the Day (SOTD) status "to some of the most unusable websites out there." As MacGregor saw it, Flash designers do not create content in a vacuum; they seek out indicators as to what quality design looks like and sounds like, and they use these examples to guide their production process:

> Just where does Macromedia think that Flash developers are looking for inspiration? Flazoom.com? We could be so lucky. Instead Flash developers look to the mothership, the creators of the Flash authoring tool for ideas on how to use Flash. Our clients look there too, wanting to see what the leading company for web development software says is the top of the line in site design. *Both developers and their clients check out the SOTD to see what new techniques are getting Macromedia's blessing.* Unfortunately, that blessing has been given all too often for fluff, glitz, and flash.... *A company cannot promote bad usability with every award it hands out and then blame the developers for their products poor reputation.* (emphasis in original)

MacGregor urged Macromedia to seek out developers who used Flash to make fast-loading, stable, usable websites that could serve as "solid examples" for other Flash designers. After all, he pointed out, Flash creators are not simply the lucky recipients of a powerful piece of software. They are partners whom Macromedia should support since Flash's bad reputation would serve neither side well. "While it is easy to say that Macromedia gave us a tool that puts bread on our table while allowing us to be the wacky, creative, designers we need to be, designers put bread on Macromedia's table by using that product in a way that puts it in high demand," he argued. "If our clients get turned off to Flash because of the reasons above, where will Macromedia be in the long run?" Rather than leave the direction of Flash to the "laws of survival," as Burgess indicated in the interview, MacGregor urged Macromedia to "take responsibility for their actions."

That December, the software giant unveiled an ambitious "usability awareness" campaign, a joint enterprise that enlisted the help of prominent Flash developers to reposition the technology and educate the Flash community about the importance of user-friendly experiences. A section of the Macromedia website was carved out as a resource center, announcing, "Macromedia is working closely with the design and development community, and usability experts to provide suggestions we think will help you create highly engaging user-focused sites." While maintaining a link to the software's original appeal (to create cool, high-impact web experiences), the usability initiative was a carefully engineered attempt to reframe the meaning of the technology. "Our goal is to empower the community to keep redefining what the Web can be and build cutting-edge sites that provide the best user experience," the usability materials explained.[54] A new tone emerged that shifted from Macromedia *versus* Flash developers to Macromedia *and* Flash developers. The collaborative spirit of "educating the Flash community" now focused on reforming the practices of a particular subset of designers: the problem was with the cottage industry of Flash enthusiasts, amateur-turned-bedroom-professionals who parlayed their skills building personal sites into paid projects. It was not the fault of a great piece of software that the less experienced became so captivated with the possibilities of Flash that they neglected to consider the end user of the website. Nor was it the fault of the experienced Flash professionals who spent years learning to harness the technology for good rather than for evil. The two sides would have to work together to teach new Flash designers how to use the software properly. An article on the web developers' resource center SitePoint, titled "All Flash: A Fast Track to Failure," made the situation explicit:

> The overall feeling of most Web professionals today is that excessive use of Flash . . . on business sites, is most definitely not the fault of Flash itself. The problems lie in the fact that Flash, like many other Web technologies, is a fairly simple tool for the "have a go" amateur to "cobble something together" with. However, dedicated Flash designers tell me that, like many Web technologies (such as scripting languages like PHP and ASP), Flash really takes years to master. Dedication, experience and an understanding of the Flash medium enable designers to use it very effectively and

efficiently to enhance a Website's overall look and feel, and communicate particular messages. Thus the "My friend knows Flash" approach for a business Website whose primary purpose is to serve information and attract customers is a killer every time.

In other words, Flash is not a "bad tool"; it just needs to be used "in the right way, preferably by a talented and experienced Flash designer."[55] Like the wild and impulsive amateur investors who first helped establish outrageous prices for internet stocks and then contributed to the drastic fall in stock values by pulling out when the going got tough, the novice Flash designer needed some schooling on the "right way" to make websites. The proper usage would be forcefully articulated in a set of mutually agreed-on usability guidelines for all Flash developers (but particularly the "have-a-go" amateur) to follow.

Macromedia's usability resource center featured "Macromedia's Top 10 Usability Tips for Flash Web Sites," which served as the centerpiece for advice on creating Flash content with a user in mind. By urging designers to "remember user goals" and "remember site goals," the tips focused on meeting user expectations and facilitating the communication needs of the client or brand. This could be achieved by avoiding "unnecessary" site intros and animation that did not reinforce site goals, tell a story, or aid in navigation. Sound should be "kept to a minimum," and navigation should always be visible and used "logically" and "consistently" to keep the user oriented at all times. Developers were urged to consider download time, provide alternative text for users with disabilities, and test their site on "someone with fresh eyes," preferably with users that matched the demographics of the site's anticipated audience.

These guidelines conveniently served two purposes. First, they provided good material for the press release that announced how Macromedia was enabling "more than 1.5 million developers using Macromedia products to create even more effective and usable Websites."[56] The usability center was billed as a "first step" that demonstrated how "the Macromedia Flash community is engaging in a conversation about the importance of user-centric design, usability testing, and accessibility." Peter Goldie, vice president of product marketing, emphasized that Macromedia is "committed to continuing to enable developers to create compelling content that provides effective user experiences."[57] Clearly,

Macromedia was *doing something* about the problems caused by the proliferation of annoying, gratuitous, self-indulgent Flash content.

Second, the guidelines left enough "interpretive flexibility" that allowed designers to maintain creative control of their web production. While "unnecessary" intros and animations should be avoided, there were plenty of projects that simply "demanded" these rich media solutions. All animation, the guidelines argued, should contribute to site goals or to storytelling or should aid navigation in some way. Sound should enhance your site but not be indispensible, the guidelines explained. But, "Macromedia's Top 10 Usability Tips" noted, "when you do use sound, Macromedia Flash will compress music into small MP3 files and even stream it." In this way, sound could even help meet the goals of tip number eight, which suggested that developers "target low-bandwidth users."

Perhaps recognizing that the new Flash enthusiasts-turned-small-scale-professionals may not be swayed by the tips provided, Macromedia also included content that traditionally attracted aspiring Flash developers: a skill-promotion contest, featured showcase sites, and inspirational quotes from Flash masters. The Design for Usability contest included categories for both "novice" and "professional" designers and rewarded winners with Apple Powerbook G4 laptops. When the winners were announced, Macromedia claimed that the 700 entries that were received from around the world proved that "designers everywhere are enthusiastic about creating usable sites with Macromedia Flash."[58] Considering the one and a half million developers who use Macromedia products, the contest in fact proved to be a very modest affair. Compared with the popular SOTD showcase center, which offered an archive of links featuring hundreds of Flash sites deemed "particularly beautiful" or demonstrating "exquisite technological engineering,"[59] the usability examples showcased included just three sites: General Motors, Land Rover, and iShopHere. Exquisite and beautiful these sites were not; instead, the GM site was commended for "good navigation" that was "clean and consistent." Land Rover "employs a nice use of animation" that reinforces navigation. And iShopHere "does a nice job of integrating animation by highlighting products" and includes a "unique feature" in the "use of search."[60] How could users *not* be inspired about usability after reading such practical descriptions?

Linking the usability initiative to Flash masters was one way Macromedia demonstrated the collaborative nature of the campaign and cued developers to read the campaign as authorized by creative voices. Peppered throughout the website were quotes from recognizable designers who offered some perspective on the importance of usability. A common trope that emerged in this discourse was the constant articulation of Flash with power. "With a tool as powerful as Flash, it's tempting to put the messenger before the message, . . . sort of like delivering a bouquet of beautiful roses with no note attached. . . . Just remember that at the core of everything we do as designers is the need to communicate," advised the noted Flash designer Hillman Curtis.[61] Josh Ulm of the experimental Remedi Project remarked that Flash was, without question, "the easiest way for developers to create powerful, emotional, compelling content for the internet." But, he suggested, there were effective and ineffective ways to use the technology: "Sound, animation, and interactivity do not have to be limited to beeps, flashes, and buttons. When used effectively Flash sites can inform, educate, and entertain. However, when it is used inappropriately, it can lead to deep frustration and even resentment."[62] These lessons were further emphasized in the new Flash development books published in 2002 that capitalized on designers teaching designers how to make usable Flash websites. All of these books forcefully articulated that there are right and wrong ways to use the technology. In *Flash: 99% Good*, for example, the authors advise, "When applied correctly, Flash can be an effective tool for engaging and retaining users. . . . When used incorrectly for things like lengthy intros, mystifying navigation, and annoying, unstoppable audio, then Flash becomes a nuisance, leaving users with a bad taste in their mouths." But in addition to acknowledging the good and the bad, an important part of the lesson involved educating designers about the arguments that usability proponents were making and teaching them how to create sites that would "stop the abuses."[63]

In "Bad Flashers Anonymous," the first chapter of *Skip Intro: Macromedia Flash Usability and Interface Design*, authors Duncan McAlester and Michelangelo Capraro acknowledge that "it's so easy to get carried away" when using a fun software program such as Flash. "You have all this power at your fingertips, how can you help but use it?" First, the authors identify what bad Flash looks like: "the irrelevant Hollywood blockbuster-style intro, the complex, drawn-out transformations triggered by a simple click of a button, the noise and the drama." Next they explain how these representational practices not only detract from the user's experience but also tarnish the reputation of the software for all developers:

> Sadly, this kind of design makes Flash and those who promote it an easy target for the knockers, and it diverts attention from all the great, effective Flash work being produced. In an era when the usability of Internet sites is such a hot topic, all that flashy Flash is a sitting duck. Imagine the fun the usability gurus have every time they come across a 2MB Flash movie on the home page of a site. Picture them gleefully switching over to 28K modem connections so that they can time how long it will take the so-called average web user to download this piece of cyber-frippery. For them it's just another Flash designer banging yet another nail in the coffin of the program they love to hate.[64]

This kind of appeal not only addresses readers as software users but invites them to see themselves as part of something much bigger: a community of Flash designers and developers who are out there creating great work. In a sense, it becomes the social responsibility of the designer to work within the boundaries of the authorized discourse to protect the greater cause and rescue the software from the grips of the gleeful, stopwatch-toting usability gurus. The first step, the authors suggest, and "possibly the hardest in the road to recovery" is to admit you have a problem. "Your heart is all for producing an award-winning stroke of genius, and your creative juices are urging you to push the boundaries of design so that your peers will just bow their heads in awe when they see your work." But we must heed the small voice inside that tells us that this just is not right; the stark truth of the matter is that "someone has to use this site. And that someone, not the people you want so dearly to impress, is who you're ultimately working for."[65] Usability design is presented as a bitter pill to swallow, but in the end, it will make the web a much healthier place.

Despite these visible efforts to turn the tide of resentment, Macromedia and the network of connected developers still struggled to rein in the excesses of Flash designers who ignored the usability guidelines. Jakob

Nielsen, who had long been blaming egotistic designers and agencies, suggested that the roots of the problem actually went much deeper. In a roundtable discussion on usability in the February 2001 issue of *Cre@te Online*, Nielsen contended,

> I do think that on the one hand it is only a tool, but on the other hand, the way the tool is done encourages problems. I think you can point the blame at the tool vendor for not having set up decent guidelines. . . . One of the reasons that Macintosh was superior to Windows for so many years was that Apple from day one had very strong human interface guidelines. They were able to beat developers over the head if they didn't follow the guidelines. The Mac software . . . had a lot of good principles built into it so that you had to do extra work to not make it good! It's always much harder to put something in after it has already escaped into the world. It's much harder to control. So I think one can blame Flash's creators for launching a tool that encourages bad design.[66]

All along, Macromedia and expert developers had been quite vocal in framing poor usability not as a fault of the tool but as the irresponsible deployment of it in the hands of the uninitiated. Even Chris MacGregor's initial defense of developers pointed the finger not at the software but at how Macromedia promoted the Flash websites it held up as exemplary. But now Nielsen was suggesting that the tool itself encouraged "bad design." The problem, in other words, was that by design, the software was presenting a preferred reading that conflicted with the usability guidelines. Even with strong social pressures exerted by prominent developers, it was difficult to intervene in established practices. Flash aesthetics were so embedded in web-based cultural production and supported by users who experienced such a significant and abiding pleasure in producing Flash-heavy content that efforts to police their practices were failing.

A Platform for Rich Internet Applications

By examining the design and evolution of the Flash application, along with the accompanying marketing and support materials, we gain some insight into how the tool and its packaging frame particular uses and interpretations as natural or intuitive. Using the application in ways similar to the demos, then, "makes sense" to designers, while "other uses and purposes seem less familiar, less likely, less viable."[67] Interface metaphors—familiar activities such as cut, copy, and paste—structure how we are permitted, encouraged, or forbidden to interact with computers, data, and media. As Lev Manovich argues, these metaphors not only shape the possibilities for interaction but also rely on a set of codes that impose a particular logic on our cultural systems and provide distinct models of the world. "Far from being a transparent window into the data inside a computer, the interface brings with it strong messages of its own."[68]

What kinds of messages did Flash send to designers, and how did these messages change over the course of the product's history? Attending to changes in software design and product positioning offers one way to glimpse broader shifts in how the web is conceived and imagined within an industrial context. Although by the late 1990s Flash was synonymous with animated websites, the software has roots in a different computing culture. In the early 1990s, "pen computing," a new generation of portable computers that allowed users to write directly on the screen with a stylus, was seen as a promising alternative to keyboard and mouse PCs. To capitalize on this new market, a company called GO Corporation developed PenPoint, an operating system for pen computers.[69] When the programmer Jonathan Gay co-founded FutureWave Software in 1993, he had hopes that his software product, SmartSketch, would dominate the market for graphics software on pen computers.[70] GO was bought by AT&T, which pulled the plug on the technology in 1994 when it became apparent that pen computing was not catching on.[71] As a result, FutureWave found itself with a product that had no market. So as not to completely scrap the work, SmartSketch was revamped first as a graphics program for Mac and Windows home computers and then as an animation tool called CelAnimator for producing animated vector graphics for the web. By summer 1996, FutureWave was learning that the market for its animation product was much broader than cel-based animation or cartoons. Beta testers were using the software to create navigational panels, technical illustrations, and advertising banners. To reflect this new positioning, the software was renamed FutureSplash Animator.[72]

Because the program was designed to make pen-based computer drawing feel more like drawing on paper (recognizing, for example, pressure-sensitive marks), the FutureSplash user interface (UI), tutori-

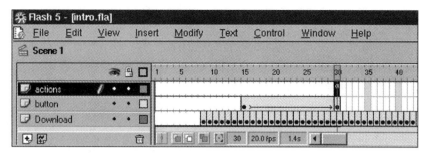

Figure 5.2. The Flash 5 interface organized production through a series of multiple overlapping timelines (2000).

als, and PR materials emphasized "fluid drawing tools" (including techniques for interacting with Bézier curves, antialiasing, and font support) along with terms drawn from traditional cel-based animation (such as "onion-skinning" and "key frames"). These utilities and metaphors were combined with language drawn from film and video editing. The FutureSplash UI referred to all of the content created with the application as "movies," even though help files described a myriad of other uses for the software (e.g., to create vector-based illustrations, printed brochures and flyers, and presentation storyboards).[73] All new projects were automatically named "Movie1" until the file was saved under a different name. Significantly, the application's foundational environment operated according to a timeline metaphor organized around the principle of the frame and the temporal state of the playhead, which moved across all available frames and "scenes" (see figure 5.2). Basic commands could be set, telling the playhead to "Stop" and "GoTo" another frame or scene when a symbol that was designated a "button" was pressed.

As the software evolved, the basic organizing principle of the timeline with its playhead, keyframes, and movie files became standardized and ingrained. New functions were added, such as the ability to include and control sound (in Version 2, 1997) and the introduction of a new "powerful" symbol called the "movie clip" (Version 3, 1998). The movie clip symbol was deemed powerful (in tutorials, help documents, software reviews) because it included a timeline of its own that operated independently of the main timeline.[74] This allowed Flash users to introduce more complex interactivity than was possible by simply controlling the position of a playhead on a single timeline. With movie clips, users could orchestrate the interactions among multiple timelines that follow different instructions and communicate with one another; this made it possible for sites to be built entirely in Flash.

Keeping track of multiple timelines (and timelines within timelines), each operating according to its own individual instructions that can "talk to" and change the playback of other timelines, however, demanded considerable mental work. To help keep it all straight, many tutorials advised users to think of themselves as directors managing a large cast of actors who need very specific instructions to perform their roles properly. In *Flash to the Core*, for example, Joshua Davis offers the following advice: "Think of your creative elements not as representational 'Symbols,' but as Objects you create that are completely in your power. My approach is that I am a director and my Objects are my 'actors.' I'm going to ask my actors to do various things on the Stage. What is most important to the principle of the Stage is that the acting, in response to my stage directions, is going to take place over time."[75] Such analogies, drawing as they do on theater, film, and television, shape the psychic and material process of development; they become not just analogies but systems of web practice that encompass tools, techniques, workflow, learning processes, and modes of sharing and communicating. They form conceptual structures that categorize what Flash is, what it means, and how it exists in relation to the web. And they make presumptions about the world in which the tool exists and the uses for which it is intended.[76] Unlike an author or a designer, who tends to work alone, a director conducts an ensemble of players, overseeing the creative or dramatic aspects of the production.

So while the "blockbuster-style intro," the "drawn-out transformations," the "noise and the drama" are not necessarily built into Flash, many of the components of the program—including the interface, the organizational metaphors, and the sample files included with the program—emphasized cinematic style and temporal production processes that lent themselves well to introductory sequences, scene transformations, and audiovisual effects that came to be derided for their lack of usability. As the software matured, more sophisticated programming capabilities were incorporated, but the foundational timeline paradigm that was developed for animation creation was initially kept in place.

In Version 4 (1999), a basic programming language called ActionScript was added, giving users even more control over the various elements of a Flash movie, including the ability to manage outside data and control variables. ActionScript was further developed in subsequent versions of Flash, adding new and more sophisticated capabilities that moved the software from an animation program to a platform for building applications, games, and mobile content. By 2002, basic animation and cartoons were only one aspect of the software's functionality, but Flash developers were still bound to the timeline system, writing ActionScript code within keyframes of the Flash movie. This was a radically different programming paradigm than the procedural or functional languages that many programmers were familiar with, and many found the sequential nature of the ActionScript programming style to be limiting, inefficient, and difficult to modify.

When the critiques of gratuitous, excessive Flash grew louder in the two years following the crash of the dot-com economy, Macromedia intensified its efforts to change the story of Flash on two visible fronts. First, to directly address usability and accessibility critiques, the company made a strategic decision to focus Flash on applications development rather than animation; the software program was substantially revised with Flash MX in March 2002 and Flash MX Professional in September 2003, and a new programming framework called Macromedia Flex was introduced in March 2004 for developing what Macromedia described as "rich internet applications" (or RIAs) on the Flash platform.[77] Second, Macromedia dealt with Nielsen's vocal critiques by hiring him as a usability consultant to develop "best practice guidelines for creating usable rich internet applications with Macromedia Flash MX."[78] Kevin Lynch, the chief software architect at Macromedia, conceded that "there are a lot of good points that Jakob has made about design usability [that have] really started some significant discussion in the Flash community, and you've seen some of that reflected in Flash MX."[79]

The decision to bring Nielsen on board had an immediate effect. On June 3, 2002, the day the partnership was publicly announced, Nielsen updated his "Flash: 99% Bad" article with an addendum noting, "Flash Now Improved":

> The version of Flash introduced in 2002 (Flash MX) has solved many of the technical usability problems in previous versions of Flash.... Macromedia has also made an important strategic decision to focus on useful

use of Flash to build Internet-based applications, as opposed to the fluffy use that was prevalent two years ago when I published the above critique of the lack of usability in most Flash designs. Macromedia is pursuing the correct strategy. They improved the Flash technology. All that remains is to get the large community of Flash developers to embrace usability principles in actual designs of Internet-based applications.[80]

Macromedia's decision to steer clear of the "fluffy use" resulted in a number of major changes to the Flash interface and development process. Now imagined as a platform for applications and content delivery, Flash MX included support for streamed video with the ability to convert a variety of formats into the new Flash Video (flv) file format. The resulting compressed video file was quite small—a 5.4 MB video could be compressed down to around 40 KB, opening the door for short-form streaming video.[81] An accessibility palette enabled authors to control features such as keyboard shortcuts and tab order and to add customized content for visually impaired people to access with screen-reader software.[82] Options were added that made bookmarking and the browser's "back" button functional, and component libraries offered uniform scrollbars and textboxes so designers did not have to build these from scratch for each project.

Perhaps most radically, the programming paradigm was expanded from timeline-based programming in keyframes to an object-oriented framework that allowed developers to code in a number of separate tools that appealed to different programming styles.[83] In short, Flash became more developer friendly, and the production process for advanced interactivity became less cinematic. One of Nielsen's major critiques was that Flash content was "typically superficial" and "tends to be created once and then left alone."[84] But as the Flash designer Tom Arah argues, "This wasn't because Flash designers are easily bored, lose interest and [are] keen to show off in pastures new. It comes down to the inherently design-intensive nature of Flash authoring. Even the simplest task of adding a new page, or rather screen, involves manually setting up frames and actions and links and adding and formatting your text is seriously laborious. Ultimately each Flash project is indeed a self-contained application, and it's just not practical to rewrite a Flash project every time new information has to be accommodated."[85] Flash MX was now capable of handling external content that could be added and changed in real-time without going through the authoring tool.

With the introduction of Flex in 2004, a developer-friendly programming environment for creating Flash-based interfaces, Flash was redesigned as a platform for building RIAs. Some of the new sites built on the Flash platform—YouTube and Flickr, for example—were attracting considerable attention. Soon these sites were among the exemplars of a whole new vision of the web, one populated by "user-generated content." These values became central to the notion of Web 2.0, a phoenix arising from the ashes of dot-com design. The ideology of Web 2.0 expressed hope for democratic participation, trust in the web's commercial potential, and a preoccupation with quantifying status and social influence online. But it was also a post-dot-com, post-9/11 phenomenon reliant on the material arrangement of already-existing technologies and the social networks of experienced practitioners and the skills, approaches, ideas, and relationships they brought with them from past work experiences. In material and semiotic terms, Web 2.0 offered a way to harness the excesses of the past in the service of a different, more egalitarian future: one in which the web might become a less sensational and more usable everyday experience for millions of ordinary users.

From Usability to UX: The Case of Adaptive Path

One year after the NASDAQ's peak in March 2000, the dot-com economy was decimated, and the market only continued its downward spiral for another year and a half. Dot-com failures were reportedly daily, chronicled in dot-bomb tell-all books and sardonic websites such as Philip Kaplan's fuckedcompany.com, created in the spring of 2000 as a "dot-com dead pool" and space for laid-off workers to share horror stories and gossip while betting on the next internet companies to go under. The dot-com pure plays, largely e-commerce companies ("e-tailers") operating exclusively over the internet, were the first to succumb to the shakeout. Their disappearance had a large impact on the internet consultancies and web services providers who counted these companies as clients and managed their internet operations. According to the Information Technology Association of America, more than half a million IT jobs in the US were lost in 2001, with technology and digital media capitals such as New York City (Silicon Alley) and the San Francisco Bay Area (Silicon Valley) among the hardest hit.[86] Selling clients on internet services during an economic downturn required a more pragmatic approach, one far removed from the swagger and vaporous promises of innovation associated with the New Economy.

Surviving internet consultancies, to signal that they too cared about "business fundamentals," rebranded around promises to put clients and customers first, emphasizing new metrics for measuring return on investment (ROI). In an ethnographic account of the new media workplace in New York's Silicon Alley, *No-Collar* (2004), Andrew Ross reports on the transformations taking place at the digital consultancy Razorfish over the course of the dot-com collapse. Describing some of the "creative frictions" he witnessed at his first team meeting in November 2000 between strategists and designers, Ross recounts a heated conversation about the company's low marks on a recent satisfaction survey. The problem, in sum, was that clients felt that Razorfish did not listen. While strategists agreed with the sentiment, designers bristled at the charge, one responding, "Experiments and risk taking are what we live for, and these clients generally don't want it. Is it our job to educate them, or does the culture not support risk taking any more? Many of these clients are learning, but they still don't understand what we do. I don't think they really appreciate that we are adding incredible value to their companies."[87] This objection is rooted in a dot-com design mentality that links value to the creative expression of coolness, attitude, and brand image necessary to win over savvy youth markets in a digital age. But by the close of 2000, there was little room for this style of thinking in internet company culture. "This is a service and relationship business. Trust equals intimacy divided by risk," the Razorfish manager reminded her team. "It won't work anymore to say to clients that we know what we're doing. We need to build this trust."[88] To survive in the twenty-first century, the e-consulting landscape would need to prioritize accountability and come up with new key performance indicators. Usability research, informed by empirical studies of actual users, could help clients feel that their websites were not just the product of some designer's creative whim but were instead business investments, "backed up by solid research on user experience."[89]

In this way, the sharp emphasis on usability and user experience extends far beyond the realm of Flash software. The charge of "unusable" captured something of the big-picture problems plaguing dot-com design. From the breathless media reports of hedonistic startups peddling unproven business models to the everyday frustrations of web designers struggling to make their sites work on both Explorer and Navigator despite the inconsistent implementation of web standards that resulted from Microsoft and Netscape's Browser Wars, the carnage of the dot-coms required a cleanup and fix-it crew. Many of those who pursued web careers during the dot-com gold rush headed back to bricks-and-mortar jobs in previous lines of employment, while more seasoned web veterans turned to freelance gigs or became independent consultants.[90]

Despite the economic uncertainty and increased competition for fewer and fewer digital accounts, there was also relief that the frenzy had finally subsided. If there was a silver lining to the bust, it was expressed through an abiding commitment to the medium and to the communities of designers and developers who stuck around to sort out the mess and fix the problems. Many web workers went solo after the mergers, layoffs, bankruptcies, and corporate restructurings had wreaked their havoc, shifting to freelance projects and consulting work. Others joined forces, pooling their resources and specializations to weather the storm. This was the strategy of Adaptive Path, which was among the first to frame "quality user experiences" as an essential competitive advantage. To understand how dot-com design was reconfigured as UX in the post-bubble climate, I focus on three significant transformations taking place across web industries after the bubble burst: the reconfiguration of the agency, the reconceptualization of design, and the fastening of new cultural imaginaries to the material production of the web.

The User-Centered Agency

By the turn of the millennium, the work of designing, marketing, and maintaining corporate websites had become a very complex undertaking. Three years earlier, a single webmaster could design and manage the whole website. But the gust of attention, interest, energy, and capital that began accumulating around the web helped the interactive services industry grow exponentially in a short amount of time. As I discussed in chapter 4, the rampant mergers and acquisitions at the height of the bubble saw nearly every successful small and medium internet company rolled into massive mega-agencies. Powerhouses such as MarchFirst, Modem Media / Poppe Tyson, and iXL aspired to become one-stop shops uniting strategy, technical, and creative services under one corporate umbrella. Industry concentration also resulted in clashes between different company cultures and work styles. The industry was still so young that there were no standard practices when it came to web development methods. "Companies had different words for similar processes. . . . Some newly acquired companies used incompatible software. We couldn't get things to work, and to make it worse, everyone came in with different system for documenting their work. It was a total mess," recalled co-founder Jesse James Garrett, who left a company that went through three acquisitions in less than a year to help start Adaptive Path.[91]

It was against this industrial climate that Garrett and six other dot-com veterans (Lane Becker, Janice Fraser, Jesse James Garrett, Mike Kuniavsky, Peter Merholz, Jeffrey Veen, and Indi Young) joined forces to start a new design consultancy in March 2001. Pooling their resources and skills, the co-founders hoped to weather the economic downturn by sharing space and clients. But this was not merely a marriage forged under dire circumstances; the principals were united around a shared design philosophy that informed the organization, strategy, and mission of the entire agency. Each of the co-founders had been working in the web industry since the mid-1990s, and together they brought a vast range of skills as speakers, writers, content developers, information architects, usability consultants, and business strategists. They authored popular books on web strategy and had become sought-after speakers at interaction design conferences. They billed themselves as "User Experience" design strategists.

Although the collaborative endeavor of well-known tech-industry "thought leaders" earned attention by virtue of name recognition, the reasons for establishing a user-centered agency were both practical and theoretical. Jeffrey Veen, a former *HotWired* designer who became Adaptive Path's director of product design, described the circumstances motivating the group's decision to start the agency: "All of the big companies spent a tremendous amount of money on these websites that didn't work. They were just baffling to people. There were giant Flash intro pages. There was all this stuff. People were like, 'That's great. How do I send my package back?' . . . So we put ourselves out as a user experience company. [Usability] was an old term that was becoming part of the vernacular in business for kind of the first time. We said, 'We will help you realize your investment you already had in your site by making it more usable.'"[92] The upside of the crash was that there were a lot of broken things about the web that needed to be fixed. In the sober socioeconomic context that followed the crash, the showmanship and spectacle that signaled quality in the era of dot-com design was radically revised.

Consider the advice and tone of the book *Heroes.com: The Names and Faces behind the Dot Com Era* (2000), a collection that profiles "a new breed of individuals—the Internet entrepreneurs who are shaping the future of the digital or 'new economy.'"[93] The timing of the book's debut was unfortunate; it hit the market at precisely the moment when the dot-com attitude was being called out as part of the problem. In the introduction, dated May 2000, the book's author, Louise Proddow, identifies a set of qualities or "attitudes" that collectively define the dot-com hero. This is an individual who "lives for today and enjoys the momentum and buzz of the Internet." Dot-com heroes are "passionate about the Internet as an agent of change." They use technology as a "competitive weapon." They are risk-takers and fast movers who play by a new set of rules. They are not afraid to "think big," and they live by the mantra to "dot com *now*."[94]

In contrast, the postcrash design philosophy exemplified by Adaptive Path was less concerned with making a splash than with making things work, designing websites that were less frustrating and easier to use. When I spoke with Veen over a decade after his days at Adaptive Path, I asked him how the idea of a "good site" changed after the bust. He replied without hesitation:

> I'll tell you one thing. One of the biggest changes that I've seen is that for even the first ten years of the web, the people who were consumers of websites, the audience of all the websites, were also the people who kind of knew how to make websites. So that gave you kind of free pass for things like usability and how things should work because you had an internal understanding. . . . Increasingly the web became just people—just people that wanted to see stuff. . . . So I think the shift has been that the people who make websites need empathy that they did not need before because the people that came to their sites were just like them. Good de-

sign has changed. Good design became, "Look what we can do to let me help you." And that's the thing, I think, that changed.[95]

"Empathy" is not among the buzzwords associated with the dot-com entrepreneur. Those interviewed in *Heroes.com* speak collectively of "putting customers first," but none explicitly mention the role of "users" in their dot-com philosophy. The postcrash design consultancy of Adaptive Path, on the other hand, was organized from the ground up around "empathy for the user." As Garrett explains,

> There's a lot about Adaptive Path that is a reaction against what we saw as being the broken practices of, specifically, consultancies. Peter [Merholz] and I both having that background felt like—we felt like they were focused on the wrong things. We felt like this user experience thing was turning into something that you could build out as a stand-alone offer rather than as an adjunct to design or technology services. So we kind of established user experience as our banner, our watchword as part of that, saying we were about this other set of things.... We start kind of pushing, really beating the UX drum far and wide, and kind of rallying people to it.[96]

Differentiating their approach from the practices of dot-com consultancies as well as the usability models of Human-Computer Interaction (HCI) specialists became a signature component of Adaptive Path's brand identity and design process. Usability experts such as Jakob Nielsen, enemy of Flash designers everywhere, made a binary distinction between two basic approaches to design: "the artistic ideal of expressing yourself and the engineering ideal of solving a problem for the customer."[97] A strong advocate of the latter, Nielsen evangelized a systematic method for usability engineering based on science, user testing, and task analysis. Usability was positioned as the flip side of graphic design approaches that emphasized expression, affect, and feeling. But for the principals of Adaptive Path, both usability and visual design were extremely limiting. Co-founder Peter Merholz explains,

> The early part of my reaction against the superficiality of graphic and visual design practice was that it wasn't respecting the medium that it was within. It was trying to apply mind-sets of prior media to this one, and it

just wasn't—it was shallow. That kept happening for the longest time. Occasionally, graphic designers would recognize it and go, "Fine, fine, fine." Then something like Flash comes along, and it's like, "Aha, here's a place where we can specify exactly how we want our things to look!" . . . I was never necessarily anti-Flash. . . . I [disagreed with] how it was being used as a means to maintain pixel-level control over the user's experience and interface, when it wasn't necessarily appropriate. . . . I was so frustrated at a graphic designer view of the world where beauty was all that mattered, when it was like, yes, it's beautiful, but no one can actually get anything done.[98]

Co-founder Mike Kuniavsky, who conducted usability testing at *Hot-Wired* before joining Adaptive Path, found usability equally problematic. "The big shift is from usability to User Experience design," he explained. "By 2001, usability had taken on this very, very reductionist notion of design. . . . User Experience design took on this larger notion that it's not just about efficiency, it's not just about accomplishing tasks. It's about creating, essentially, a narrative of somebody's experience in somebody's mind about what it is that they're doing."[99]

The new discourse of "User Experience," which quickly gained traction between 2002 and 2005 within web design and HCI communities, combined insights of user-centered design methodologies from library and information sciences with lessons from Flash design's appeal to "rich web experiences." The former recognizes the task-oriented nature of using software; the latter emphasizes the emotional texture and sensory components of an experience. "We start thinking about these things," says Garrett. "Visiting a website is not just about accomplishing a goal: How does it make people feel? What's the visual texture of the thing?"[100]

The assemblage of User Experience design (UX) was accomplished through the merging of distinct development communities that began attending conferences together, codifying practices, and devising a nomenclature to help large, diverse teams identify project roles across disciplines. The creation of a shared "contact language" was crucial for the web design's reconfiguration in the twenty-first century.[101] The Adaptive Path co-founders became active participants and facilitators of public discourse about User Experience design: in addition to client work, the company began hosting annual conferences, including its flagship event, UX Week, which became a key annual event in the interaction design industry.

By 2005, the name Adaptive Path had become almost synonymous with "User Experience." The industry was developing a new set of "best practices," values, approaches, and design methodologies that offered a new way of measuring a website's return on investment (ROI) on the basis of User Experience as a tool for delivering business value. This was not the same discourse of experience that sought to "wow" visitors with Flash experiences; nor was it a discourse of usability that shunned aesthetics in favor of utility and functionality above all else. Instead, UX understood that aesthetic experiences play a crucial role in users' perception of value. Desire and emotion, in other words, were no longer antithetical to the idea of usable design. Engaging users, inviting them into websites as participants and creators, became a core idea that was articulated to new visions of the web, new models for measuring ROI, and new technical infrastructures for supporting what Web 2.0 devotees called "architectures of participation." Collaborative projects such as Wikipedia, the rise of peer-to-peer file sharing in the wake of Napster, and the popularity of new platforms (such as YouTube and Flickr) for sharing user-generated content were held up as celebrated examples of a "whole new web." In the wake of the stock-market crash and 9/11, dot-com design—as an industrial practice, economic system, rhetorical strategy, and belief system—was no longer a useful, usable social imaginary but a threatening obstruction that blocked the view to a better interactive future. This "better" future promised to overwrite the mistakes of the dot-com era by linking user participation and sharing to democracy. User Experience and Web 2.0 congealed as a powerful discursive formation that gained momentum by serving the needs of multiple constituents (interaction designers, investors, tech journalists, consultants, start-ups, etc.) at work on different types of problems. Why would internet veterans at this time *not* find something compelling about this discourse? The reconfiguration of dot-com design into a new assemblage of User Experience offered a second chance to showcase the meaningfulness of the web and to reinvent a collective imaginary around network culture that might better accommodate the everyday practices of ordinary users.

Conclusion

Reconfiguring Web Histories

March 12, 2014, marked the "official" twenty-fifth anniversary of the World Wide Web. The highly anticipated occasion was promoted by the international web standards-setting consortium, W3C, and the World Wide Web Foundation, which celebrated by launching an anniversary website (www.webat25.org) urging visitors to share their web memories and tweet birthday greetings using the hashtag #web25. Hailed as an opportunity to reflect on the history and future of the web, the event was perhaps equally remarkable for revealing both the web's profound significance as an everyday media and information technology and the overwhelming confusion of many users in understanding what, precisely, was being celebrated. Some tweeted greetings using a new hashtag, #Happy25thInternet, while others seemed to assume that on this date twenty-five years ago, the web suddenly appeared fully formed and ready to use.

Of course, what is lost in this celebration of anniversaries and inventions is the complicated work of dating a birth: twenty-five years earlier, there was not a single web page online. There were no browsers. The protocols for making the web functional had yet to be written. Instead, what the official anniversary commemorated was the first attempt by Tim Berners-Lee, the web's primary inventor, to articulate his vision of a shared universal information space. Written in March 1989, "Information Management: A Proposal," offered a rough blueprint of a technical and imagined hyperspace—a designed environment—that was originally conceived as a document-management system to help keep track of all of the small pieces, people, and equipment that make up large projects under way at CERN, the Swiss physics research institute where Berners-Lee worked. Although he struggled for roughly five more years to convince others of the project's potential, by the end of the 1990s, the

World Wide Web had thoroughly captivated the popular imagination of the US public, financial investors, and the mainstream media.

Today's web is very different from the one Berners-Lee imagined in 1989. Does it even matter that most people wishing it a happy birthday had no idea that they were celebrating a proposal (a rejected one, at that!) for global hypertext? Or that the internet and the World Wide Web are actually different things? For everyday users of the network just going about life and work, it really does not matter—that is actually one of the gifts that today's internet users inherited from the countless known and unknown individuals who toiled on its design. The rough and uncharted "frontier" web of the early to mid-1990s has been smoothed over, reconfigured, commercialized, and reimagined—again and again—leaving some practices and components behind while carrying others along (for example, monetization schemes, standards, assessment criteria, and user-friendly improvements, as well as symbolically resonant narratives, metaphors, ideologies, lingering affect, and desires) into the next iteration and another assemblage. Dot-com design has been deterritorialized, and a new formation has gathered momentum: User Experience design.

But for those who care about these networked media and communication environments, who build, maintain, and study them, knowledge of past efforts, struggles, beliefs, failures, and assumptions is crucial. The past offers resources for thinking differently about the present, inspiring us to wonder how it is that things got to be the way they are, helping us question those assumptions that came to be known as common sense. By paying closer attention to the ways the web was historically imagined, designed, and continually revised, we might better understand today's web, apps, and social media platforms as an outcome of a series of struggles that took place over the course of the web's first decade. There is a popular saying about the importance of learning history or else risk being condemned to repeat the same mistakes of the past. Yet there are examples presented in this book that describe moments when practices once deemed mistakes (or mis-takes) later do take hold in a different context, time, or place and become manifest as another assemblage that achieves hegemony.

In chapter 3, for example, I described two related "internet spectaculars"—ambitious global projects designed to put the power of the

World Wide Web on display—launched in 1995 and 1996. The earlier website, MIT's *A Day in the Life of Cyberspace*, proffered a model of the web as a platform, powered by the contributions of users who logged in to access their capsules, individual profile pages that were automatically updated as users participated, chatted, and played online. When this ambitious collaboration between the MIT Media Lab and Art Technology Group was launched in the fall of 1995, the interactive features were a novelty, but this model of "social" media failed to take off at the time. Instead, it was the following project, *24 Hours in Cyberspace*, that received more attention, praise, and publicity. This website was technically ambitious too, but instead of directing the technical infrastructure toward an architecture of participation that supports the contributions of users, the whole project was designed to coordinate the social networks of media professionals who were globally collaborating behind the scenes. The resulting *24 Hours in Cyberspace* website, composed of web pages rendered on the fly from preformatted templates that created stylistic uniformity and a magazine aesthetic, was a celebrated mainstream-media event that was covered by every broadcast network on the evening news.

Undoubtedly, many users of the time had never seen a website that looked so polished and packaged. It borrowed organizational and layout conventions from print and used those markers to signal professional-quality design, a sentiment that found agreement among a host of users and web designers looking to establish some professional authority within a new discipline and medium. Later critics snidely disparaged this style and mode of production as the read-only web. What we have lost sight of, however, are the reasons why so many web designers and users found agreement on this model of design at the time. Programmers and engineers bemoaned the incorrect use of HTML tags by graphic designers trying to gain some visual control over a technical, low-res medium. But for those who have been trained in print, the work of visual arrangement and order is what makes information feel easier to digest, clear, and understandable—it is readable because it is formatted in familiar ways. Bringing traditional graphic design to the web was a way of inviting a mass audience online. Masking the complexity of the system and making the interface feel intuitive has helped the internet transform from a network that only specialists could use to one that is (ostensibly) designed for everyone.

Conceptualizing Design

Let us return, then, to one of the key problematics that this book takes up: what exactly do we mean by "design"? More than just the subject matter addressed across five chapters, design is understood in this book as a mode of inquiry that encompasses both idea conceptualization and the finished product or realized outcome of these resources. Often paired with another term to describe a particular creative practice ("graphic design," "costume design," "interior design," etc.), "design" is often understood as style, aesthetics, and visual appearance. But in a range of other disciplines from science and technology studies (STS) to anthropology, "design" has emerged as a provocative keyword that challenges traditional assumptions and methodologies.

As both a verb and a noun, "design" encompasses the plan, process, and product; it references not only a way of making or creating things but also the process of designing—the relational work of arranging and fitting together. As a term that blends theory and practice together, "design" shares with process philosophy an orientation toward relational approaches that emphasize the dynamic process of "becoming" through change and transformation, rather than focusing on fixed or static sites of "being." This way of thinking about design has much in common with Deleuze's concept of assemblage (*agencement* in French), which means "layout, put together, combine, fitting."[1] It refers to the tentative fitting together and unfolding of heterogeneous components that cohere for a time, "constructed in part as they are entangled together."[2] Like "design," "assemblage" retains the duel meaning of both the arrangement of things and the act of arranging.

Design depends on carving out boundaries; it regulates by giving form and order to things. Each chapter of this book approaches the concept and practice of design from multiple perspectives and specializations: programmers, engineers, amateurs, information architects, graphic artists, systems integrators, and even users are all regarded as designers. Herbert Spencer famously described this broader conceptualization of design with the assertion, "Everyone designs who devises courses of action aimed at changing existing conditions into preferred ones."[3] One of the existing conditions that this book aims to challenge is the sedimentation of the web's historical periodization as split into Web 1.0 and Web

2.0 eras. The latter term, "Web 2.0," coined by O'Reilly Media in 2003, has largely fallen out of use in the internet technology and startup world where it originated, having been replaced by the more marketable "social media." (Try asking today's college students if they know what Web 2.0 is.) The industry has moved on, and so have the ordinary users who have integrated the internet into all areas of everyday life.

Instead, it seems that one of the strongholds of the version model of web history can be found in academic scholarship, where boundaries dividing Web 1.0 and Web 2.0 have already been ingrained and codified. While many scholars have critiqued the ideological construction of Web 2.0 as an industrial discourse that trades on promises of democracy and participation, as a historical marker, the boundary still holds. One of the reasons why this boundary is simultaneously undermined and reinforced in internet research and media studies is likely because these scholars care about making sense of contemporary practices and need to contextualize changes in attitudes, modes of production, technologies, and user practices. The discourse of versions offers a built-in boundary marker that makes it easier to map these developments against a stable past to illuminate the present. Web 2.0 literally creates a contemporary context in which there is something new to study. But how might attention to design help us interrogate and reconfigure boundaries?

Design and Production in Media Studies

Over the past few decades, design studies has shifted its focus from the object or designed artifact to understand design more broadly as a historically situated mode of inquiry, one that is bound up with imagining the future. Today, the phrase "design thinking"—figured as a user-centered, prototype-driven process of problem solving—has become a buzzword in scientific, engineering, and industrial sectors concerned with innovation. Meanwhile, "speculative design" approaches try to challenge normative assumptions about innovation by designing prototypes that imagine alternative or more equitable futures. Thinking about design, then, is an invitation to remap boundaries and divisions between objects, concepts, agencies, people, time, and space.

This work of boundary interrogation draws on feminist technoscience studies that question the orders, categories, and divisions that are often taken to be natural—binary gender or sexuality categories, for example. But this approach also offers a critical vantage point that would be useful in critical media studies. As a conceptual framework, design offers media studies a chance to expand the way we talk about production. For decades, media studies has been organized around a production-text-reception model that was more suited to a mass-media environment than to the hybrid mass-personal-social forms that accompanied the growth of the internet. Thinking of media as designed and not just produced may open new ways of conceiving objects of studies and redrawing boundaries to ask new questions. The separation of production and consumption was a key move in cultural studies initiated by Stuart Hall's seminal 1973 paper "Encoding and Decoding in the Television Discourse," which offered a model of media studies that accounted for the varying ways audience members read a program by arguing that there is no symmetrical or causal link between encoding (production) and decoding (consumption). For Hall and members of the Birmingham School, this was an important theoretical and political move that offered a different approach than the predominant mainstream mass-media effects models of communication. Rather than focusing on the powerful effects that mass media have on individuals, Hall's model remapped the boundaries between production and consumption by returning agency to audience members who do not all interpret texts the same way.

Hall's production-text-audience model was enormously influential in the formation of media and cultural studies and has been built on, challenged, and reworked by numerous media scholars over the decades. Yet the boundaries that Hall drew to intervene in mass-media models in the 1970s are still in place today, even though many media scholars have pointed to the limits of dividing production from consumption in the age of digital media and global information networks. Media and technology companies such as Apple, Amazon, and Google, for instance, have shifted their approaches from designing stand-alone devices (a phone, a television) to designing relations and contexts using new vocabularies (e.g., "architectures of feedback," "participatory design"). We might see products such as the smart phone, internet-enabled television, or lifestyle apps as both the conception and the result of reconfigured boundaries between bodies, data, systems, things, and content. If media and technological convergence has altered the way that designers (professionals and everyday users alike) understand, experience, and produce mediated environments today, then as media scholars, we should think carefully about how we conceive our objects of study and design our own research projects (what is "outside the bounds of this study").

Of course, design as a critical framework is not without drawbacks. As Lucy Suchman points out, design solutions offers a shiny veneer to the problems of the world; they present the world as always in need of design.[4] Similarly, as Lily Irani and her collaborators on the design of a Mechanical Turk project realized, design carries connotations, and they are not always positive ones.[5] Their project was designed to raise awareness of imbalances and inequalities, but their identity as designers frustrated their efforts to deliver on their message. Design comes with certain class-based connotations: design is Target, not Walmart; design is Ikea, not Sears. Design and production both signal class-based work. Designers, "the creative class," are inventive problem solvers. Production, on the other hand, retains a link to media studies' indebtedness to Marxism and the power struggles between the owners of the means of production and the proletariat workers. Although this divide between production and consumption is an artifact of an industrial economy, the concern with uneven power dynamics, class struggle, and ideology are still relevant in an information age.

Designing Historical Methods

In the introduction, I asked readers to keep in mind that this is not just a book about design but also a designed book. By now, I hope readers understand that by calling attention to this book's design, I am not referring to the cover art, marketing materials, typesetting, or illustrations. Designing a book involves packaging ideas and pitching them to a publisher. It involves constructing a narrative that makes sense and is supported with evidence, choosing examples and structuring chapter breaks. In constructing the conceptual and material design of *Dot-Com Design*, I chose to configure the past against the economic lines of the NASDAQ composite stock index. By breaking chapters along the stages of a classic speculative bubble, however, I risk the chance that readers may think I am implying a causal link between stock-market activity and web aesthetics. However, it is not my intention to reinforce an underlying economic determinism or to suggest that changes in the market caused changes in web design practices (or vice versa).

Rather, I want to emphasize that the lived experience of working on the web in the 1990s was colored against the backdrop of the market—it was always there, measuring and quantifying mood and translating it into value. To render this background context of "the market" (which is both a real and imagined phenomenon) more visible, to bring out the passion and pessimism that framed the backdrop of working life in the 1990s, I use stock-market activity as a structural device for organizing the boundaries of each chapter. Throughout the 1990s, Andrew Ross argues, American workers became more and more aware of how market manipulation affected their jobs and financial security, but nowhere was this more evident than in the publicly traded dot-coms. He explains:

> In companies where options were standard issue, it had become customary to put a stockwatch application on the desktop screen of all the computers. As a result, employees got used to a daily round of emotional highs and lows that were directly tied to the fluctuations in the company's stock price. . . . In publicly traded New Economy companies, the feature was extended to production-level employees, thereby distributing the psychology of risk downward. It was an unusual form of motivation, and it gave rise to a new species of industrial anxiety. Where once the working day had been dictated by the regimen of the factory clock, now it was regulated by the flux of the stock index.[6]

Making the market more visible was part of dot-com design, and so it is part of *Dot-Com Design*. But drawing attention to finance capital and the market is also a way to preserve something valuable from media and cultural studies traditions, including political economy and media industry studies, and that is attunement to the power structures and class relations that undergird the means and relations of production.

The web is always "under construction," even as it moves beyond personal computers and browsers into the realm of mobile devices, stand-alone applications, and internet-enabled high-definition television sets. By tracking shifts in web practice—the connected system of creative workers, new media industries, rules and assumptions, work processes and methodologies, software, "best practice" guidelines, aesthetics, and stylistic forms—I have shown how the "look of the web" is not merely the result of technologies that inaugurate new creative practices or business decisions (foolish or wise) that structure the work of web production. Nor is the postcrash pursuit of user-centered design, and the subsequent emergence of a user-generated Web 2.0, a simple case of web designers coming to their senses at last. Instead, it is a register of a new dominant discourse that gained momentum in response to shifts in the larger socioeconomic context.

Revisiting the early web is not just an exercise in nostalgia but a way to illuminate a whole ensemble of connections encompassing aesthetics, ideologies, economies, and industries. As we move forward to analyze the new media and technologies that populate the mobile web, the networked television screen, the app store, and the tablet computer, we would do well to see these new logics as not unconnected to the power struggles of the recent past. By historically grounding the constellation of people, practices, technology, institutions, and social imaginaries that came together as an assemblage that I have called dot-com design, I have mapped alternative routes through web history that avoid the teleological progress narratives that accompany Web 1.0 and Web 2.0 discourses. This book concentrates on the US context because this is where web advertising and commercial internet investments got started. The dot-com bubble, however, was not just a US phenomenon, and the formation of professional web development industries took place around the world. Studying these industries requires close attention to the specific circumstances and cultural contexts of design. If we hope to have a better grasp of how global networks have been configured and imagined in local situated contexts, we will need to rely on many more accounts that can take us beyond the dominant culture of dot-com design. I hope this book initiates more historical work in this direction. Indeed, there are many more web histories that have yet to be told.

ACKNOWLEDGMENTS

This project has been as long in the making as the time span it covers. It began before it was a history! Many of the driving questions that animate this book have their genesis in my experiences working as a web designer and teaching digital media classes in Baltimore, Maryland, between 1997 and 2002.

I owe my start down this path to the good fortune of encountering Dr. Elliot King at Loyola University Maryland, my former professor who first took me under his wing, taught me HTML, and helped me find a new direction in postcollege life—initially as a teaching assistant for his "Introduction to Internet Publishing" course and then as an adjunct instructor and research associate at Loyola's New Media Center. My years working with Elliot in the classroom, community, and lab during the dot-com boom were invigorating; I fondly recall mapping the architecture of websites on whiteboards together, chatting about technology, religion, and the stock market while CNBC supplied a constant stream of stock quotes and market updates in the background. Elliot first inspired me to think critically about digital culture and web aesthetics, and this book would not exist if it were not for his intellectual generosity and boundless energy for starting new projects and bringing newbies like me on board.

Teaching principles of interaction design over this span of six years, a period encompassing the very height of dot-com exuberance and the fallout after the market crashed in 2000, I was struck by how often the emerging conventions defining "good" web design or "best practices" changed. Every semester I searched out and passed along web design tips and guidelines, "how-to" tutorials, examples of "good" websites, and lists of the "Top 10 biggest web design mistakes" to help keep students from making those dreadful errors found on the "What *Not* to Do" lists. And nearly every semester the rules seemed to change. Should websites be silent? Should they look like magazine spreads or television commercials? Should users click through content, or should they scroll? The web was a brand-new medium in the 1990s. Who was to decide what it should look and feel like? These were the questions that motivated my research, and I am indebted to the many mentors and friends I met at the University of Wisconsin–Madison who helped me formulate the shape of this project.

A number of people played a key role in my intellectual and professional development. My advisor, Michael Curtin, nurtured my interest in cultural industries and helped me carve a critical path through the world of commercial web development. Michele Hilmes, who makes everyone want to be a historian, brought me into the archives and taught me how to think historically. Jonathan Gray has been a constant source of encouragement and a generous advocate of my work. Mary Beltrán, Julie D'Acci, Greg Downey, Rob Howard, Shanti Kumar, and Eric Schatzberg have all offered valuable feedback at various stages along the way. Lisa Nakamura was the reason I chose to attend grad school at Madison in the first place; the years we spent working together on the development of her "Critical Internet Studies" course has had an enormous influence on this project, as have her seminars on visual culture, race, and the internet. She has been a constant source of support and encouragement, first through my graduate years at Madison and now as a mentor, friend, and colleague at Michigan, where she helped transform our small cadre of internet recluses from different departments into a kick-ass home for Digital Studies. In research, service, and in life, her model of grit and resilience continues to inspire.

I have benefited from the care and mentorship of amazing colleagues at the University of Michigan who have been vital sources of support and community. Aswin Punathambekar has been an indispensible resource, and I am personally grateful to have him as a friend, colleague, mentor, and writing buddy. Amanda Lotz has gone above and beyond many times over in her mentorship duties—providing comments on this book, career and teaching advice, and thought-provoking conversations over cocktails. Susan Douglas has always been willing to listen, share advice, and offer professional and emotional support. Katherine Sender has kindly tolerated my lengthy ramblings and usually helps me get to the point. Nicole Ellison keeps me afloat with her friendship and has given me a whole new appreciation for data points. It has been a pleasure to participate in the stimulating cross-disciplinary environment of Digital Studies at Michigan and a privilege to hang out with smart, interesting people such as Tung Hui-Hu, John Cheney-Lippold, Anna Watkins Fisher, Sheila Murphy, Sarah Murray, Christian Sandvig, Melanie Yergeau, and Irina Aristarkhova. I have also benefited from numerous conversations with my colleagues in Communication Studies, Screen Arts, and beyond—many thanks go to Andre Brock, Scott Campbell, Robin Means Coleman, Sonya Dal Cin, Kris Harrison, Muzamil Hussein, Shazia Ithkar, Yeidy Rivero, Michael Schudson, Julia Sonnevend, and Derek Vaillant. Finally, I am deeply indebted to Paddy Scannell, who helped me find my footing with this project at exactly the right moment when he had me over for tea and asked, "How does a website become a useful, usable thing?" Just when I thought all was lost, he introduced me to the notion of care structures in theory and in practice. I am grateful not only for the personal care structure he provided by being available but also for opening my eyes to the invisible environment of hidden concern that is materialized in the world around us.

A number of individuals read this manuscript during different stages of development and offered incisive feedback that helped me clarify the main argument and narrative. I am grateful to Tara McPherson and Fred Turner for taking the time to read my work and for asking tough questions early on. The Social Media Collective group at Microsoft Research generously supported my visit to Cambridge for a week, and I am grateful to Nancy Baym, Jean Burgess, Kate Crawford, Megan Finn, Mary Gray, and Kate Miltner for the hospitality and for providing helpful feedback on chapter 2. I have also been part of an amazing writing group for the past five years and could not begin to express the heartfelt appreciation for the thoughtful comments and perspectives of those who have been involved at one time or another. Charlotte Karem Albrecht, Dan Herbert, Hui-Hui Hu, Sarah Murray, Aswin Punathambekar, Katherine Sender, Greg Schneider-Bateman, and Pavitra Sundar are all brilliant and kind; one could simply not ask for a better group to learn from.

Many others have been important resources in both formal and informal ways. I thank Eric Zinner and Lisha Nadkarni at NYU Press and series editors Nina Huntemann and Jonathan Gray for all of their help, time, and patience. Huge thanks go to Finn Brunton, Lily Irani, Molly Wright Steenson, Michael Stevenson, Anne Helmond, Kevin Driscoll, Rob Gehl, Liz Ellcessor, Danny Kimball, and Germaine Halegoua for inspiring conversations and endless geeking out about design, technology, and web history. My fellow co-editors of the journal *Internet Histories*—Niels Brugger, Valerie Schaffer, and Gerard Goggin—all deserve standing ovations for staying with me during a crazy and difficult year. I am so appreciative of the countless conversations and insights gained over the years thanks to more friends and colleagues than I can possibly name. I want to extend a warm thanks to Ben Aslinger, Megan Biddinger, Chris Cwynar, Lindsey Hogan, Josh Jackson, Derek Johnson, Jeremy Morris, Caryn Murphy, and Erin Copple Smith.

This project was strengthened by interviews I conducted with web designers, copywriters, architects, and consultants who agreed to talk with me and share materials from their early web work. Megan Finn, Elizabeth Goodman, and Molly Wright Steenson were key intermediaries who helped me make contacts and secure interviews. In particular, I wish to thank Brian Behlendorf, Sasha Cavender, Tom Cunniff, Judith Donath, Dave Fry, Jesse James Garrett, Craig Grannell, Heather Hesketh, Jelane Johnson, Mike Kuniavsky, Charles Marelli, Peter Merholz, Matthew Nelson, Matt Owens, Jen Robbins, Alicia Rockmore, and Jeffrey Veen. Although it is impossible to tell everyone's story in a single book, my conversations with former web workers made it all the more clear that there are many more histories yet to be written.

Finally, it is to my family that I owe the most heartfelt thanks. My mother, Susan Reese, has from the very beginning always encouraged me to pursue a life of thinking. She has borne the brunt of my busy schedule and deserves a phone call every single day. My father passed away in 1997 just as I was beginning my short-lived stint in web design. Although he expressed profound trepidation about this career move at the time, he would no doubt be pleased to see that I ended up on a career path that offers health insurance (at least for now). Ingrid Ankerson saw me through the first half, and I will always remain profoundly grateful for everything she has done in support of my career and life. I think fondly of all of our past collaborations and look forward to our continued partnership as parents to two amazing kids. Sarah Murray helped me actually get this project out the door by reminding me what I care about. Her intellect and generosity have made new things possible, and I look forward to all the future may bring. Finally, my biggest gratitude goes to the two people who have sacrificed the most in the process of bringing this book to fruition. I love you Clyde Sapnar Ankerson and Harvey Sapnar Ankerson. Thanks for always boosting my spirits by telling me that I am good at technology and video games.

NOTES

INTRODUCTION

1. "1994: Today: What Is Internet, Anyway," YouTube video, accessed January 8, 2018, http://www.youtube.com/watch?v=UlJku_CSyNg.
2. Dovey and Kennedy, *Game Cultures*.
3. Grossman, "You—Yes, You—Are *Time*'s Person of the Year."
4. Marwick, *Status Update*.
5. Allen, "What Was Web 2.0?," 261.
6. Shiller, *Irrational Exuberance*; Kelly, *New Rules for the New Economy*.
7. Thrift, "It's the Romance, Not the Finance," 414.
8. Cooper, Dimitrov, and Rau, "Rose.Com by Any Other Name."
9. Foucault, *Archaeology of Knowledge*, 49.
10. Neff, *Venture Labor*, 11.
11. Simon, "Dot-Com Kids."
12. Deleuze and Guattari, *Thousand Plateaus*.
13. One important exception is Helen Kennedy's study of the ethical inflections of web-design workers involved in the web-standards movement. See Kennedy, *Net Work*.
14. Phillips, "Agencement/Assemblage"; Wise, "Assemblage"; Callon, "Elaborating the Notion of Performativity"; Puar, "I Would Rather Be a Cyborg than a Goddess."
15. Phillips, "Agencement/Assemblage," 108.
16. Dwiggins, *Layout in Advertising*, 190, quoted in Remington, *American Modernism*, 41.
17. Barnard, *Graphic Design as Communication*, 128.
18. Spigel, "Object Lessons for the Media Home," 538.
19. Buchanan and Margolin, *Idea of Design*.
20. Suchman, *Plans and Situated Actions*, 77.
21. Appadurai, *Future as Cultural Fact*, 254.
22. Brown, *Change by Design*; Lockwood, *Design Thinking*; Martin, *Design of Business*; Cross, *Design Thinking*.
23. Kimbell, "Rethinking Design Thinking," 286.
24. Ibid.
25. Buchanan, "Design and the New Rhetoric," 186.
26. See, for example, Nussbaum, "Design Thinking Is a Failed Experiment, So What's Next?"
27. Suchman, "Anthropological Relocations and the Limits of Design," 5.

28. Dunne and Raby, *Speculative Everything*, 34.
29. Ibid., 2.
30. DiSalvo, "Spectacles and Tropes," 110.
31. Appadurai, *Modernity at Large*, 7.
32. Ibid.
33. Ibid., 8.
34. Ibid., 7.
35. Kapor and Barlow, "Across the Electronic Frontier."
36. Kelty, *Two Bits*, 38.
37. Preda, *Framing Finance*; Stäheli, *Spectacular Speculation*.
38. Preda's *Framing Finance* is premised on identifying how groups associated with stock exchanges in the nineteenth century managed to redefine finance as a scientific pursuit grounded in "observational boundaries," systems through which the stock exchange represents itself to the public. The book is, therefore, completely invested in examining how technologies of seeing and observing are crucial to the maintenance of financial systems. However, Preda does not explicitly reflect on the relationship between speculation and seeing.
39. Preda, *Framing Finance*, 86.
40. Lefèvre, *Traité des valeurs mobilières*, 245, quoted in Preda, "Informative Prices, Rational Investors," 359.
41. Preda, *Framing Finance*, 78.
42. Callens, *Les maîtres de l'erreur*, 270–72, cited in Jovanovic, "Economic Instruments and Theory," 180.
43. Lefèvre, *Traité des valeurs mobilières*, 184–85, quoted in Preda, "Informative Prices, Rational Investors," 372.
44. Jovanovic, "Economic Instruments and Theory," 169.
45. Ibid., 184.
46. Ibid., 180.
47. Preda, *Framing Finance*, 89.
48. Knafo, "Financial Crises and the Political Economy of Speculative Bubbles."
49. Preda, *Framing Finance*, 92.
50. Hewitson, *Feminist Economics*.
51. Stäheli, *Spectacular Speculation*, 171–72.
52. Newman and Levine, *Legitimating Television*, 6.
53. For research that investigates these types of questions, see Josephson, Barnes, and Lipton, *Visualizing the Web*.
54. Shah and Kesan, "Privatization of the Internet's Backbone Network."
55. Aspray and Ceruzzi, *Internet and American Business*.
56. Turow, *Niche Envy*.
57. Kennedy, *Net Work*.
58. Internet Archive WayBack Machine home page, https://archive.org/web.
59. Kindleberger, *Manias, Panics, and Crashes*, 13.
60. Ibid.

CHAPTER 1. FORGING A NEW MEDIA IMAGINATION (1991–1994)

Epigraph: Berners-Lee, *Weaving the Web*, 27.

1. Abbate, *Inventing the Internet*, 2.
2. Kindleberger, *Manias, Panics, and Crashes*, 27.
3. Cassidy, *Dot.Con*, 4.
4. Hall and Massey, "Interpreting the Crisis," 57.
5. Hecht, "Rupture Talk in the Nuclear Age."
6. Gillies and Cailliau, *How the Web Was Born*, 136–41.
7. *Understanding the Internet*.
8. Berners-Lee, *Weaving the Web*, 10–12.
9. Ibid., 37.
10. Frana, "Before the Web There Was Gopher," 25.
11. Ibid., 29.
12. Ibid.
13. Mark McCahill, UseNet post to comp.infosystems.gopher, March 10, 1993, quoted ibid., 30.
14. Berners-Lee, *Weaving the Web*, 73.
15. Ibid., 74.
16. Frana, "Before the Web There Was Gopher," 33.
17. Berners-Lee, *Weaving the Web*, 16.
18. Ibid., 15.
19. Ibid., 20.
20. T. Nelson, "Complex Information Processing," 85.
21. Bardini, *Bootstrapping*.
22. See, for example, Bush, Nyce, and Kahn, *From Memex to Hypertext*; Conklin, "Hypertext."
23. Bush, "As We May Think," 106.
24. Berners-Lee, *Weaving the Web*, 6.
25. Ibid., 16.
26. Berners-Lee, *Weaving the Web*, 39. (Today, URIs comprise two subclasses: URLs and Uniform Resource Names, or URNs, which reference persistent, location-independent identifiers for resources.)
27. While Gopher servers could be bookmarked, the Gopher client stores the server name and path as separate pieces of information, which means they were not "addressable" and could not easily be shared, on, say, a billboard or a box of cereal. See L. Richardson, "System of the World Wide Web."
28. Liu, *Laws of Cool*, 145.
29. Berners-Lee, *Weaving the Web*, 72; Frana, "Before the Web There Was Gopher," 34.
30. Berners-Lee, *Weaving the Web*, 26.
31. Ibid., 27.
32. Lyotard, *Postmodern Condition*, xxiv.
33. Bolter, *Writing Space*, 135; Landow, *Hypertext 2.0*, 89.

34. Landow and Delany, "Hypertext, Hypermedia and Literary Studies," 6.
35. Frana, "Before the Web There Was Gopher," 21. Frana cites Turkle, *Life on the Screen*, 17.
36. Terranova, *Network Culture*, 47.
37. Ibid., 56.
38. Ibid., 1.
39. Grossberg, "On Postmodernism and Articulation," 47.
40. Jameson, *Postmodernism*, 43; D. Harvey, *Condition of Postmodernity*, 189.
41. Frank, *One Market under God*, xiv.
42. Gillies and Cailliau, *How the Web Was Born*, 240.
43. Berners-Lee, *Weaving the Web*, 68.
44. Berners-Lee et al., "World-Wide Web," 80.
45. Streeter, *Net Effect*, 126.
46. Ibid., 127.
47. Streeter, "Moment of *Wired*," 770.
48. "New Poll Confirms Internet as Information Superhighway."
49. Freedom Forum Media Studies Center, *Special Report*.
50. Clinton and Gore, *Putting People First*, 77.
51. Ceruzzi, *History of Modern Computing*, 322.
52. Information Infrastructure Task Force, *National Information Infrastructure*, 6.
53. Markoff, "Building the Electronic Superhighway." See also Streeter, *Net Effect*, 106–24; Flichy, *Internet Imaginaire*, 17–34.
54. Elmer-Dewitt and Jackson, "Take a Trip into the Future."
55. Wolff, *Burn Rate*, 116.
56. Ibid.
57. An edited transcript from an interview with Barry Diller on *Larry King Live*, April 12, 1993, was reprinted in *Advertising Age*: "Shopping on the Interactive Highway," 22.
58. Elmer-Dewitt, "Take a Trip into the Future," 52.
59. Solomon et al., "Dueling Dudes."
60. Sacks and Maher, "Wall Street Wires Up the New World of Media," 14.
61. Ibid.
62. Quoted in Burstein and Kline, *Road Warriors*, 272 (italics in original).
63. Gibbons, "Marriage Made on Information Highway Annulled."
64. "Computers Seen Beating TV as Infopike Route of Choice," 10.
65. Streeter, *Net Effect*, 123.
66. Levy, "How the Propeller Heads Stole the Electronic Future," quoted in Flichy, *Internet Imaginaire*, 32.
67. *ABC Evening News with Peter Jennings*, TV broadcast, ABC, September 30, 1993.
68. Burstein and Kline, *Road Warriors*, 126–27.
69. Ibid., 104.
70. Ibid., 127.
71. Streeter, *Net Effect*, 123–24.

72. Turner, *From Counterculture to Cyberculture*, 142.
73. Gibson, *Neuromancer*, 67.
74. Stone, "Will the Real Body Please Stand Up?," 98.
75. In response, Gibson threatened to trademark the name of Eric Gullichsen, the lead programmer of the Autodesk Cyberspace project. The company ended up trademarking "Autodesk Cyberspace" instead.
76. See Turner, *From Counterculture to Cyberculture*, 141–74.
77. Gibson, "Cyberspace '90," 107.
78. Greenwald, "*Wired* 01.01." This oral history accompanied the reissued first edition of *Wired* for the iPad, which included interviews and commentary.
79. Rossetto, "Story of *Wired*."
80. Greenwald, "*Wired* 01.01."
81. Turner, *From Counterculture to Cyberculture*, 212.
82. Schwartz, "Propeller Head Heaven," 62.
83. Keegan, "Reality Distortion Field."
84. Wolf, *Wired*, 31.
85. Greenwald, "*Wired* 01.01."
86. Ibid.
87. For more on the role of media formats in signaling how people fit within different identities and lifestyles, see Wolf, *Wired*, 34.
88. Dear, "Revisiting the Original 1992 *Wired* Media Kit."
89. Ibid.
90. Keegan, "Digerati!"
91. Greenwald, "*Wired* 01.01."
92. Turner, *From Counterculture to Cyberculture*, 73.
93. Heller, "Reputations."
94. See Rothstein, "*Wired* Magazine in the Nineties," 111.
95. The first three years of these opening spreads were published by HardWired, *Wired*'s book division, in 1996. See McLuhan and Fiore, *Medium Is the Massage*.
96. Plunkett and Rossetto, *Mind Grenades*.
97. González-Vélez, "Assessing the Conceptual Use of Social Imagination."
98. Streeter, *Net Effect*, 124–25.
99. Appadurai, *Modernity at Large*, 31.

CHAPTER 2. COOL QUALITY AND THE COMMERCIAL WEB (1994–1995)

1. CERN, "Birth of the Web."
2. Moody, "Internet."
3. Berners-Lee, *Weaving the Web*, 47.
4. See Gillies and Cailliau, *How the Web Was Born*, 245; WWW Virtual Library, "WWW Virtual Library."
5. The original page hosted at NCSA is no longer online, but numerous copies of "What's New with NCSA Mosaic" can still be found online. See, for example, http://www.w3.org/MarkUp/html-test/ncsa/whats-new.html.

6. Mosaic Communications Corporation, "What's New."
7. Rosenberg, *Say Everything*, 10.
8. The First International Conference on the World-Wide Web took place in Geneva, Switzerland, in May 1994; see First International Conference on the World-Wide Web, "WWW94—Awards."
9. Bruns, "From Homepages to Network Profiles," 418.
10. Nielsen, "HCI and the Web."
11. D. Kaplan, *Silicon Boys*, 304–7.
12. Hahn and Stout, *Internet Yellow Pages*; Jamsa and Cope, *World Wide Web Directory*.
13. Moody, "Internet."
14. R. Richardson, *Web.Guide*.
15. Turlington, *Walking the World Wide Web*.
16. See, for example, Wolff et al., *Netguide*; Renehan, *1001 Really Cool Web Sites*.
17. Brunsdon, "Problems with Quality," 67.
18. Ibid., 73.
19. Lemay, *Teach Yourself Web Publishing*, 212.
20. Liu, *Laws of Cool*, 306.
21. Ibid., 78.
22. Ibid., 186–93.
23. Ibid., 77–78 (italics in original).
24. For distinctions between "big chronologies" and "small chronologies," see Edwards et al., "AHR Conversation."
25. The courts ruled against Apple because the company had already granted Microsoft permission to use and sublicense some elements of its visual displays for Windows 1.0. Apple argued that Windows 2.0, Windows 3.0, and Microsoft's licensee, Hewlett-Packard, had exceeded the terms of the license because these operating systems had "striking similarities in the animation of overlapping windows and the design, layout and animation of icons," thus making Windows more "Mac-like." *Apple Computer, Inc. v. Microsoft Corp.*, 35 F.3d 1435, 1442–45, 32 U.S.P.Q2d 1086 (9th Cir. 1994). For analysis of trade dress protection and the look and feel of graphical interfaces, see Terry, "Gui Wars"; Kellner, "Trade Dress Protection."
26. Mirzoeff, "On Visuality," 67.
27. Mirzoeff, *Right to Look*, 3.
28. Williams, *Marxism and Literature*, 131.
29. Ibid., 132–33.
30. The term "anthologies of cool" is from Liu, *Laws of Cool*, 186.
31. Renehan, *1001 Really Cool Web Sites*; D. Taylor, *Creating Cool Web Pages with HTML*; *CyberHound's Internet Guide*.
32. The original website was http://www.infi.net/cool.html.
33. Ryan, "What's Cool on Line?"
34. G. Davis, "Announce."

35. The Cool Site of the Day archive is available through the WayBack Machine at http://web.archive.org/web/19980118171814/http://www.coolsiteoftheday.com/cgi-bin/stillcool.pl?month=08&year=94.
36. Liu, *Laws of Cool*, 187.
37. Frankel, "Word & Image."
38. Wolf, "Justin in the Zone."
39. McPherson, "Reload," 462.
40. Ibid., 462.
41. "What Is Coolium?"
42. Quoted in Liu, *Laws of Cool*, 455n7.
43. See, for example, "Vadim's Hotlist as of 9/11/94," http://www.vmt-com.com/hotlist.html.
44. Renehan, *1001 Really Cool Web Sites*, vi.
45. R. Richardson, *Web.Guide*, 2.
46. This website is still available at http://www.het.brown.edu/people/mende/benjamin.
47. Wolf, "(Second Phase of the) Revolution Has Begun" (italics in original).
48. Williams, *Marxism and Literature*, 133–34.
49. *CyberHound's Internet Guide*, vii.
50. Smith and King, "Hard Sell."
51. Yavovich, "Burnett's Interactive Crystal Ball," 16.
52. Emmrich, "Agencies Prepare Clients."
53. Artzt, "P&G's Artzt," 24.
54. Ibid.
55. McMillan, "Death of Advertising May Be Premature," 212.
56. Brunton, *Spam*, 56.
57. Lewis, "Sneering at a Virtual Lynch Mob."
58. Canter and Siegel, *How to Make a Fortune on the Information Superhighway*.
59. Donaton, "Ketchum Communications," 17.
60. D'Angelo, "Happy Birthday, Digital Advertising!"
61. Ibid.
62. Ibid.
63. See Treffiletti, "Rick Boyce," 7–8.
64. D'Angelo, "Happy Birthday, Digital Advertising!"
65. Routledge, "Organic Online."
66. C. Taylor, "Recruiting for the Interactive Age," 12.
67. Bayers, "Original Internet Adman," 36.
68. Treffiletti, "Rick Boyce," 8.
69. Cleland, "Upstart Company Snares Big Internet Marketing Gigs," 24.
70. Cleland, "Fear Creates Strange Bedfellows," 16.
71. Ibid.
72. C. Taylor, "Ordering Interactive a La Carte," 29.
73. R. Cook, "Behind the Hype," *Campaign*, 37.

74. Evenson, "Wired to Go Online," E1.
75. Marchand, *Advertising the American Dream*, 87.
76. Ibid., 90.
77. Smulyan, *Selling Radio*, 69.
78. Marchand, *Advertising the American Dream*, 90.
79. Quoted in Smulyen, *Selling Radio*, 71.
80. Even after the NSF relinquished control of the internet backbone, "netiquette" still remained, and therefore so did brochureware. Since the first commercial sites helped to set a standard of what a commercial website was supposed to look like, brochureware persisted, as smaller companies on much more modest budgets borrowed the formal structure of these first sites. Today the term is still widely used, often to critique static, HTML-only websites that do not provide social networking or e-commerce.
81. Lotz, "Textual (Im)Possibilities," 39.
82. "Club Med Rockets along the Information Highway."
83. Turow, *Breaking Up America*, 125.
84. Berkowitz, "Marketers Explore Cyberspace."
85. Charles Marelli, interview with author, March 10, 2011.
86. "High Stakes in Cyberspace."
87. Marelli, interview.
88. Thanks to Phil Gyford, who once worked for the UK version of *Wired* magazine and provided me with a 1995 copy of the *HotWired* site and selections from Duncan's episodes on Zima.com.
89. C. Taylor, "Z Factor," 15.
90. Ibid., 14.
91. H. Lewis, "Copywriting for Interactive Media," 238.
92. Marelli, "Anatomy of a Web Advertisement."
93. Ibid.
94. C. Taylor, "Z Factor," 16.
95. Williamson, "Latest Pitch," 13.
96. C. Taylor, "Mama Says," 26.
97. Renehan, *1001 Really Cool Web Sites*, 794.
98. C. Taylor, "Mama Says," 26.
99. Tom Cunniff, interview with author, April 12, 2010.
100. C. Taylor, "Mama Says," 26.
101. Fry Multimedia, "Website Design for Ragu."
102. These sound clips were the voice of Antonio Antiochia, a software engineer with Fry Multimedia.
103. Standage, "Menu Masters," 17.
104. The site could also be accessed at www.ragu.com.
105. Wolf, *Wired*, 138.
106. Scott, "Suds Marketers Look to Future."
107. "Molson Adds Personal E-mail to Its Site," 19.

108. Borland, "Browser Wars."
109. Carroll, Broadhead, and Cassell, *Canadian Internet Handbook*, 53.
110. When the site launched in July 1995, both the "Cooler" and "Canadian Culture" sections were listed as "under construction."
111. Scott, "Suds Marketers Look to Future."
112. Ibid.
113. Lyons, "Lack of Hard Numbers Fails to Deter Stampede," 61.

CHAPTER 3. DESIGNING A WEB OF LEGITIMATE EXPERTS (1995–1998)

1. For critiques of these online-only approaches, see Slater, "Social Relationships"; Markham and Baym, *Internet Inquiry*; Papacharissi, "Virtual Sphere."
2. Slater, "Social Relationships," 553.
3. See Parrika, *What Is Media Archaeology?*; Chun, *Programmed Visions*; Montfort and Bogost, *Racing the Beam*; Kirschenbaum, *Mechanisms*.
4. Parikka, *What Is Media Archaeology?*, 83.
5. Wardrip-Fruin, "Digital Media Archaeology," 302.
6. O'Reilly, "Web 2.0 Compact Definition."
7. Turner, *From Counterculture to Cyberculture*, 72.
8. Girard and Stark, "Heterarchies of Value," 80.
9. Ibid., 97.
10. Barthes, *Mythologies*, xi.
11. Marwick, *Status Update*.
12. Allen, "What Was Web 2.0?"
13. Gillmor, *We the Media*.
14. Stevenson, "Rethinking the Participatory Web."
15. O'Reilly, "Web 2.0 Compact Definition."
16. Couldry and van Dijck, "Researching Social Media as If the Social Mattered."
17. Livingstone, *Audiences and Publics*.
18. Benkler, *Wealth of Networks*; Papacharissi, "Virtual Sphere"; boyd, "Why Youth (Heart) Social Network Sites."
19. Thompson, *Media and Modernity*; Butsch, *Media and Public Spheres*.
20. Law, *After Method*, 84–85.
21. Ibid., 41–42.
22. Suchman, "Configuration."
23. Ibid., 50, quoting Daston, "Coming into Being of Scientific Objects," 1.
24. Warner, *Publics and Counterpublics*, 90.
25. Anderson, *Imagined Communities*.
26. Jones and Rafaeli, "Time to Split, Virtually."
27. Benkler, *Wealth of Networks*.
28. Marwick and boyd, "I Tweet Honestly, I Tweet Passionately."
29. MIT Media Lab, "1010 Revisited."
30. Donath, "Inhabiting the Virtual City," sec. 7.1.
31. MIT Media Lab, "About This Event."

32. MIT Media Lab, "About the Book."
33. W. Grossman, "Futurology," B4.
34. MIT Media Lab, "About the Book."
35. Auerbach, "Cyberspace Spawns Fight," 53; Rifkin, "Arguing over the Dawn of Cyberspace's 'Day' Book."
36. Rifkin, "Arguing over the Dawn of Cyberspace's 'Day' Book."
37. Ibid.
38. Brand, *Media Lab*, 150.
39. Sasha Cavender, personal communication, April 19, 2014.
40. Lynch, "Flash a Smile Icon for Net Snapshot."
41. Weise, "'Day in the Life of' Crew Takes on Cyberspace."
42. Reid, "Day in the Cyberlife."
43. See Altheide and Snow, *Media Logic*; van Dijck and Poell, "Understanding Social Media Logics."
44. Rifkin, "Arguing over the Dawn of Cyberspace's 'Day' Book."
45. Cavender, "On-Line World Miffed over Cyberspace 'Day' Project."
46. Streeter *Net Effect*, 124.
47. Cassidy, *Dot.Con*, 85.
48. P. Lewis, "Sneering at a Virtual Lynch Mob."
49. Warner, *Publics and Counterpublics*.
50. Scannell, "For-Anyone-as-Someone Structures," 12.
51. Donath, "Inhabiting the Virtual City," sec. 7.1.3.
52. Ibid.
53. Lacey, *Listening Publics*.
54. Weltevrede, Helmond, and Gerlitz, "Politics of Real-Time"; Hu, "Real Time/Zero Time."
55. Warner, *Publics and Counterpublics*, 97.
56. Ibid., 90.
57. MIT Media Lab, "1010 Revisited."
58. Van Dijck and Poell, "Understanding Social Media Logic." 3.
59. Parks, *Cultures in Orbit*, 33.
60. Ibid., 37.
61. Dayan and Katz, *Media Events*, 124.
62. Cline and Walter, "24 Hours in Cyberspace."
63. Lemonick, "Net's Strange Day," 55.
64. Mok, *Designing Business*, 172.
65. Ibid., 160.
66. Ankerson, "Web Industries, Economies, Aesthetics."
67. Donath, "Establishment of Identity in Virtual Communities."
68. "Art Technology Group," 1245.
69. Mok, "Netobjects."
70. Copilevitz, "How Much Is That Web Site in the Windows?," 1C.
71. Ibid.

CHAPTER 4. E-COMMERCE EUPHORIA AND AUTEURS OF THE NEW ECONOMY (1998–2000)
1. Wolff, *Burn Rate*, 133.
2. Ibid., 123.
3. Ibid., 244.
4. Ibid., 239.
5. Ibid., 55.
6. Foucault, *Archaeology of Knowledge*, 49.
7. Foucault, "Confessions of the Flesh," 194.
8. Drucker, *Landmarks of Tomorrow*.
9. Reagan, "Remarks and a Question-and-Answer Session."
10. Ibid.
11. Clinton, "Remarks of Governor Bill Clinton."
12. Wines, "Clinton Focusing on the World Economy," A10.
13. Shiller, *Irrational Exuberance*, 96.
14. Ibid., 98–99.
15. Greenspan, "Remarks by Chairman Alan Greenspan."
16. Madrick, *Business Media and the New Economy*.
17. Mandel, "Triumph of the New Economy," 68.
18. "Assembling the New Economy," 71.
19. Cassidy, *Dot.Con*, 160.
20. Shepard, "New Economy," 38.
21. Mackay, *Extraordinary Popular Delusions*.
22. Kindleberger, *Manias, Panics, and Crashes*, 15.
23. See Cassidy, *Dot.Con*, 125–27.
24. Fuerbringer, "Ruble Throws Another Scare into Markets."
25. Greenspan, *Age of Turbulence*, 191.
26. Lowenstein, *When Genius Failed*, xix.
27. Greenspan, *Age of Turbulence*, 194.
28. Morgenson, "Seeing a Fund as Too Big to Fail."
29. Brenner, *Boom and the Bubble*, 173.
30. In trading terms, a put option is a contract in which the buyer of a certain security acquires the right to sell at a pre-agreed price if the price drops.
31. Cassidy, *Dot.Con*, 189.
32. Browning, Ip, and Sicsm, "What Correction?," A1.
33. Taffler and Tuckett, "Psychoanalytic Interpretation of Dot.com Stock Valuations."
34. Schultz and Zaman, "Do the Individuals Closest to Internet Firms Believe They Are Overvalued?," 348.
35. "Technology Sector Outlook"; "If at First You Don't Succeed, Try a '.Com,'"
36. Schurr, "To Explain Web Mania," B1.
37. See Mary Meeker's comments in Bary, "'Net Queen."
38. Bartiromo and Francis, "Internet-Based Retailer CyberShop."

39. Bennett, "Online Retailers Report Revenue Grows Strongly," sec. B, 9C.
40. Meeker, *U.S. and the Americas Investment Research*, 1.
41. Laderman and Smith, "Internet Stocks," 121.
42. Shaun Andrikopoulos, an internet stock analyst for the investment bank BT Alex Brown, came up with TEMA, which projects future earnings by estimating revenue growth and the operating margins that the company would hope to achieve when it has matured.
43. Streitfeld, "High-Tech Anxiety Hits Silicon Valley," A01.
44. Gannon, "From Content to Commerce."
45. Nelson, "From Bedroom to Boardroom," 160.
46. Messina, "Crunch Time in Alley."
47. "iXL Consolidates Acquisitions."
48. Maddox, "IXL-Los Angeles Continues on Agency Buying Spree," 14.
49. Ibid.
50. Blankenhorn, "Big Time," 25.
51. "iXL Launches New Interactive Web Site."
52. "Carl Levine Joins iXL's New York Office."
53. Doyle, "Masters of the e-Channel," 46.
54. Bell, "Agencies," 24.
55. Kirsner, "Art of the I-deal," 30.
56. Rafter, "Keeping the Peace."
57. Ibid.
58. Ibid.
59. Heydebrand and Mirón, "Constructing Innovativeness in New-Media Start-Up Firms," 1962.
60. Matt Owens of the small New York web boutique onenine, interview with author, February 20, 2008.
61. Benjamin, "Net Service," A6.
62. Miège, *Capitalization of Cultural Production*, 44.
63. See Netscape, "Netscape Builds Momentum."
64. Roback, "FutureSplash Animator Makes Impressive Debut," 37.
65. Waldron, "Flash History."
66. The software company Adobe acquired Macromedia in 2005 for $3.4 billion.
67. Zeldman, "Where Have All the Designers Gone?"
68. Bitmapped images such as the GIF and JPEG images embedded in HTML pages are pixel-based formats that define colored pixels on the screen. Vector graphics rely on mathematical algorithms to render the image "on the fly," allowing them to scale without image degradation.
69. Moock, foreword to *Flash Hacks*, x.
70. Zeldman, "Where Have All of the Designers Gone?"
71. Dougherty, "Why Flash Is Significant."
72. Ibid.
73. Lunenfeld, "19?? to 20??."

74. Ibid.
75. The ActionScript programming language that was developed for Flash could be used to program images with mathematical algorithms that can be used to produce various physics-related effects. See Salter and Murray, *Flash*, 55–56.
76. Burgoyne and Faber, *New Internet Design Project*, 51.
77. Shepter, *Personal Web Sites*, 152.
78. Na, "Judge's Remarks."
79. Retrospective email interviews with eight attendees of the Flash Forward 2000 conference that was held in New York City.
80. Georgenes, "Flash Designer Gets Hooked on Community."
81. Zeldman, "Style vs. Design."
82. Dougherty, "Why Flash Is Significant."
83. Miège, *Capitalization of Cultural Production*, 46.
84. Jankowsi et al., *New Masters of Flash*, 511–25.
85. J. Davis, *PrayStation Hardrive*.
86. Lee, "Designer's Republic," 138.
87. "Window on the New Economy."
88. J. Davis, Q&A following "Creating Complexity from Simplicity."
89. See Curtin and Streeter, "Media," 233.
90. Miège, "Logics at Work."
91. Cook, "Auteur Cinema and the 'Film Generation'"; Lewis, *Hollywood v. Hard Core*; King, "From Auteurs to Brats."
92. Thompson and Burns, *Making Television*; Newcomb and Alley, *Producer's Medium*; Caldwell, *Production Culture*.

CHAPTER 5. USERS, USABILITY, AND USER EXPERIENCE (2000–2005)

1. Hershey, "Another Sell-Off in Technology," A1.
2. Ibid.
3. "Market's Breathtaking Plunge," A16.
4. Cassidy, *Dot.Con*, 299.
5. The article points out that the stock market fell almost 20 percent in 1998 and 36 percent in 1987, "but the economy chugged on." Ip and Browning, "Getting Real," A1.
6. The most common explanations involved overextended investors who had to sell to pay capital gains taxes and meet margin calls. Krugman, "Roller-Coaster Markets," A23.
7. Chernow, "Market in Need of a Broker," sec. 4, 15.
8. Ibid.
9. McGee and McGough, "Some Bulls Make a Case for Tech Stocks," C1.
10. Ibid.
11. Pender, "Analysts See US Stock Plunge," 45.
12. Cassidy, *Dot.Con*, 197.

13. Andrew Ross's book *No-Collar: The Humane Workplace and Its Hidden Costs* offers an ethnography of the internet consultancy Razorfish, which chronicles the company's and employees' response to the bursting of the internet bubble.
14. See Neff, *Venture Labor*.
15. Mackay and Gillespie, "Extending the Social Shaping of Technology Approach," 698.
16. Eglash, "Appropriating Technology," xi.
17. The term "interpretive flexibility" is a key component of Trevor Pinch and Wiebe Bijker's Social Construction of Technology (SCOT) approach, where it is used to demonstrate the negotiations between "relevant social groups" regarding the design, development, use, or meaning of a technology. See Pinch and Bijker, "Social Construction of Facts," 421. Steve Woolgar challenges the "finality" of technological artifacts, pointing out that changes in the understanding, interpretation, and description of a technology go on long after "closure." He uses the term "interpretive flexibility" to describe the different readings that users can pursue when engaging with technology. See Woolgar, "Technologies as Cultural Artifacts."
18. Woolgar, "Technologies as Cultural Artifacts."
19. Akrich, "De-Scription of Technical Objects."
20. Liu, *Laws of Cool*, 171.
21. Woolgar, "Technologies as Cultural Artifacts," 92.
22. Ibid., 90.
23. Gillespie, "Stories Digital Tools Tell," 114.
24. For example, see Leggett, de Boer, and Janousek, *Foundation Flash Applications for Mobile Devices*, 4.
25. Doull, comment on "Golden Age of Flash."
26. Dougherty, "Why Flash Is Significant."
27. See the launch issue of *Cre@te Online: The Web Designer's Bible* (July 2000).
28. This was determined by reading the descriptions that accompany each showcase, which detail how the site works and the software used to create it.
29. Cirillo, "User Advocate or Enemy of Creativity?"
30. F. Harvey, "Survey—Creative Business."
31. Nielsen, "End of Web Design."
32. Cirillo, "User Advocate Or Enemy of Creativity?"
33. Ibid.
34. Walton, "Web of Visions."
35. Paul, "Flashing the Web"; "That Mess on Your Web Site."
36. Ragus, "Flash Is Evil."
37. MacGregor, "Cancer on the Web Called Flash."
38. Yacco Vijn, interview with author, July 2008.
39. Morrison, "Focus—Usability," 2.
40. Ibid.
41. "Re: FLASH: FLASH 99% BAD."
42. Ibid.

43. Walton, "Web of Visions."
44. "Re: FLASH: FLASH 99% BAD."
45. Heid, "Call to Action."
46. Oelling, Krisher, and Holsten, "Declination of Independence."
47. Ibid.
48. Weingarten, "Flash Backlash"; Teague, "Backlash against Flash."
49. J. Davis, *Flash to the Core*, 13–14.
50. Dreamless was first launched in 1999, but after succumbing to disputes and flame wars, it was closed in July 2001.
51. Holland, "Rob Burgess."
52. Winner, "Do Artifacts Have Politics?"
53. Aymer and Dredge, "Flash Forward," 65.
54. Macromedia, "Macromedia—Flash Usability."
55. Wilson, "All-Flash: A Fast Track to Failure."
56. Macromedia, "Macromedia Launches Usability Initiative."
57. Ibid.
58. Macromedia, "Macromedia Flash 5." Submissions were accepted between February 19 and March 30, 2001. "Novice" designers were described as "individuals who have used Macromedia Flash for no more than 60 days" and "professional" designers were defined as those "who have used Macromedia Flash for more than 60 days."
59. Macromedia, "Showcase Submission Guidelines."
60. Macromedia, "Macromedia Flash: Usability Examples."
61. Macromedia, "Macromedia—Flash Usability."
62. Macromedia, "Macromedia Flash: Usability Quotes."
63. Airgid and Reindel, *Flash: 99% Good*, 5.
64. McAlester and Capraro, *Skip Intro*, 10.
65. Ibid., 11.
66. "Usability Blues," 73.
67. Gillespie, "Stories Digital Tools Tell," 112.
68. Manovich, *Language of New Media*, 65.
69. J. Kaplan, *Startup*, 36.
70. Gay, "History of Flash."
71. Matsumoto, "AT&T Pulls Plug," 4.
72. "FutureWave Software Renames CelAnimator."
73. Milburn and Warner, *Flash 2 Web Animation Book*, 11–18.
74. Plant, *Flash 3!*, 148.
75. J. Davis, *Flash to the Core*, 24.
76. Gillespie, "Stories Digital Tools Tell," 115.
77. "Internet Week—Developers Want Flash Flexibility," 20.
78. Macromedia, "Macromedia and Usability Guru Jakob Nielsen."
79. Becker, "Web Design Guru Sees Flash Challenges."
80. This news update was originally available as an addendum to Jakob Nielsen's article: "Flash: 99% Bad, Update." The update was removed and replaced with Flash

usability results after the Nielsen Norman Group completed its study of Flash for Macromedia. However, a copy of the original article can be found at http://www.mmcis.com/Resources/Articles/flash-bad.html.
81. Green, "Rise of Flash Video."
82. Arah, "Flash MX."
83. Babcock, "Macromedia Tools Move beyond Animation."
84. Nielsen, "Flash: 99% Bad."
85. Arah, "Flash MX."
86. O'Brien, "High-Tech Slowdown May Be Worsening," 1.
87. Ross, *No-Collar*, 62.
88. Ibid., 63.
89. Ibid., 226.
90. Stellin, "Sorting Out a Future after Dot-Com," 1.
91. Jesse James Garrett, interview with author, May 1, 2013.
92. Jeffrey Veen, interview with author, May 21, 2013.
93. Proddow, *Heroes.com*, foreword.
94. Ibid., introduction.
95. Veen, interview.
96. Garrett, interview.
97. Nielsen, *Designing Web Usability*, 11.
98. Peter Merholz, interview with author, April 30, 2013.
99. Mike Kuniavsky, interview with author, May 6, 2013.
100. Garrett, interview.
101. Galison, *Image and Logic*.

CONCLUSION

1. Phillips, "Agencement/Assemblage"; Wise, "Assemblage"; Callon, "Elaborating the Notion of Performativity"; Puar, "I Would Rather Be a Cyborg than a Goddess."
2. Law, *After Method*, 41.
3. H. Simon, *Sciences of the Artificial*, 111.
4. Suchman, "Anthropological Relocations," 3.
5. Irani and Silberman, "Stories We Tell about Labor."
6. Ross, *No-Collar*, 200.

BIBLIOGRAPHY

Abbate, Janet. *Inventing the Internet*. Cambridge, MA: MIT Press, 2000.

Airgid, Kevin, and Stephanie Reindel. *Flash: 99% Good: A Guide to Macromedia Flash™ Usability*. New York: McGraw-Hill, 2002.

Akrich, Madeleine. "The De-Scription of Technical Objects." In *Shaping Technology / Building Society: Studies in Sociotechnical Change*, edited by Wiebe Bijker and John Law, 205–24. Cambridge, MA: MIT Press, 1992.

Allen, Matthew. "What Was Web 2.0? Versions as the Dominant Mode of Internet History." *New Media & Society* 15, no. 2 (2013): 260–75.

Altheide, David L., and Robert P. Snow. *Media Logic*. Beverly Hills, CA: Sage, 1979.

Anderson, Benedict. *Imagined Communities: Reflections on the Origin and Spread of Nationalism*. London: Verso, 1991.

Ankerson, Megan Sapnar. "Web Industries, Economies, Aesthetics: Mapping the Look of the Web in the Dot-Com Era." In *Web History*, edited by Niels Brügger, 173–94. New York: Peter Lang, 2010.

Appadurai, Arjun. *The Future as Cultural Fact: Essays on the Global Condition*. London: Verso, 2013.

———. *Modernity at Large: Cultural Dimensions of Globalization*. Minneapolis: University of Minnesota Press, 1996.

Arah, Tom. "Flash MX—New Improved Flash?" Designer-Info.com, December 2002. http://designer-info.com/Writing/flash_usability.htm.

"Art Technology Group Enables Sun Microsystems to Personalize, Unify Enterprise-Wide E-Commerce Site." *Business Wire*, September 21, 1998, 1245.

Artzt, Ed. "P&G's Artzt: TV Advertising in Danger; Remedy Is to Embrace Technology and Return to Program Ownership." Transcript of speech. *Advertising Age*, May 23, 1994, 24.

Aspray, William, and Paul E. Ceruzzi. *The Internet and American Business*. Cambridge, MA: MIT Press, 2008.

"Assembling the New Economy." *Economist*, September 13, 1997, 71–73.

Auerbach, Jon. "Cyberspace Spawns Fight: MIT Lab, Photographer Feud over Idea for Internet Story." *Boston Globe*, January 18, 1996, 53.

Aymer, Graeme, and Stuart Dredge. "Flash Forward." *Cre@te Online* 15 (July 2001): 65.

Babcock, Charles. "Macromedia Tools Move beyond Animation." *Information Week*, August 27, 2003. http://www.informationweek.com/news/software/enterpriseapps/showArticle.jhtml?articleID=14100035.

Bardini, Thierry. *Bootstrapping: Douglas Engelbart, Coevolution, and the Origins of Personal Computing*. Stanford, CA: Stanford University Press, 2000.

Barlow, John Perry. *A Declaration of the Independence of Cyberspace*. EFF, February 9, 1996. http://homes.eff.org/~barlow/Declaration-Final.html.

———. "I'm John Perry Barlow, co-founder of EFF and Freedom of the Press Foundation. I wrote the Declaration of Independence of Cyberspace 20 years ago today." Reddit AMA, February 8, 2016. https://www.reddit.com/r/tabled/comments/44tkkg/table_iama_im_john_perry_barlow_cofounder_of_eff/.

Barnard, Malcolm. *Graphic Design as Communication*. New York: Routledge, 2013.

Barthes, Roland. *Mythologies*. 2nd ed. New York: Hill and Wang, 2013.

Bartiromo, Maria, and Bruce Francis. "Internet-Based Retailer CyberShop Ends Its First Day of Trading Up More than 30 Percent." News transcript. *Business Center*, CNBC, March 23 1998.

Bary, Andrew. "'Net Queen: How Mary Meeker Came to Rule the Internet." *Barron's*, December 21, 1998, 23–28.

Bayers, Chip. "The Original Internet Adman." *Ad Week* 52 (2011): 34–37.

Becker, David. "Web Design Guru Sees Flash Challenges." CNET, December 9, 2002. http://news.cnet.com/web-design-guru-sees-Flash-challenges/2100-1040_3-976586.html.

Bell, Daniel. *The Coming of Post-Industrial Society: A Venture in Social Forecasting*. New York: Basic Books, 1973.

Bell, John. "Agencies: One Size No Longer Fits All." *Revolutions*, March 1, 2000, 24.

Benjamin, Matthew. "Net Service: Where Suits Meet Orange-Hair Crew." *Investor's Business Daily*, August 23, 1999, A6.

Benkler, Yochai. *The Wealth of Networks: How Social Production Transforms Markets and Freedom*. New Haven, CT: Yale University Press, 2006.

Bennett, Johanna. "Online Retailers Report Revenue Grows Strongly." *Wall Street Journal*, January 12, 1999, sec. B, 9C.

Berkowitz, Harry. "Marketers Explore Cyberspace." *Newsday*, April 10, 1995.

Berners-Lee, Tim. "Information Management: A Proposal." No. CERN-DD-89-001-OC. CERN, March 1989, May 1990. http://cds.cern.ch/record/369245/files/dd-89-001.pdf.

———. *Weaving the Web: The Original Design and Ultimate Destiny of the World Wide Web*. San Francisco: HarperCollins, 2000.

Berners-Lee, Tim, Robert Cailliau, Ari Luotonen, Henrik Frystyk Nielsen, and Arthur Secret. "The World-Wide Web." *Communications of the ACM* 37, no. 8 (1994): 76–82.

Blankenhorn, Dana. "The Big Time: Ex-TV Mogul Bert Ellis Building New Fortune Online." *Business Marketing*, June 1, 1998, 25.

Bolter, J. David. *Writing Space: The Computer, Hypertext, and the History of Writing*. Hillsdale, NJ: Lawrence Erlbaum, 1991.

Borland, John. "Browser Wars: High Price, Huge Rewards." *ZDNet News*, April 15, 2003. http://news.zdnet.com/2100-3513_22-128738.html.

Bourdieu, Pierre. *Distinction: A Social Critique of the Judgement of Taste*. Translated by Richard Nice. Cambridge, MA: Harvard University Press, 1984.
boyd, danah. "Why Youth (Heart) Social Network Sites: The Role of Networked Publics in Teenage Social Life." In *Youth, Identity, and Digital Media*, edited by David Buckingham, 119–42. Cambridge, MA: MIT Press, 2008.
Brand, Stewart. *The Media Lab: Inventing the Future at MIT*. New York: Penguin, 1988.
Brenner, Robert. *The Boom and the Bubble: The U.S. In the World Economy*. London: Verso, 2002.
Brown, Tim. *Change by Design: How Design Thinking Transforms Organizations and Inspires Innovation*. New York: Harper Business, 2009.
Browning, E. S., Greg Ip, and Leslie Scism. "What Correction? With Dazzling Speed, Market Roars Back to Another New High." *Wall Street Journal*, November 24, 1998, A1.
Bruns, Axel. "From Homepages to Network Profiles: Balancing Personal and Social Identity." In *A Companion to New Media Dynamics*, edited by John Hartley, Jean Burgess, and Axel Bruns, 417–26. Malden, MA: Wiley, 2013.
Brunsdon, Charlotte. "Problems with Quality." *Screen* 31, no. 1 (1990): 67–91.
Brunton, Finn. *Spam: A Shadow History of the Internet*. Cambridge, MA: MIT Press, 2013.
Buchanan, Richard. "Design and the New Rhetoric: Productive Arts in the Philosophy of Culture." *Philosophy and Rhetoric* 34, no. 3 (2001): 183–206.
Buchanan, Richard, and Victor Margolin. *The Idea of Design: A Design Issues Reader*. Cambridge, MA: MIT Press, 2000.
Burgoyne, Patrick, and Liz Faber. *The New Internet Design Project: Reloaded*. New York: Universe, 1999.
Burstein, Daniel, and David Kline. *Road Warriors: Dreams and Nightmares Along the Information Superhighway*. New York: Dutton, 1995.
Bush, Vannevar. "As We May Think." *Atlantic Monthly* 176 (1945): 101–8.
Bush, Vannevar, James M. Nyce, and Paul Kahn. *From Memex to Hypertext: Vannevar Bush and the Mind's Machine*. Boston: Academic, 1991.
Butsch, Richard. *Media and Public Spheres*. New York: Palgrave Macmillan, 2007.
Caldwell, John Thornton. *Production Culture: Industrial Reflexivity and Critical Practice in Film and Television*. Durham, NC: Duke University Press, 2008.
Callens, Stephane. *Les maîtres de l'erreur* [Masters of error]. Paris: PUF, 1997.
Callon, Michel. "Elaborating the Notion of Performativity." *Le Libellio d'Aegis* 5, no. 1 (2009): 18–29.
Canter, Laurence A., and Martha S. Siegel. *How to Make a Fortune on the Information Superhighway*. New York: HarperCollins, 1995.
"Carl Levine Joins iXL's New York Office as Senior Business Development Manager: iXL Continues to Attract Industry's Top Talent." *PR Newswire*, December 14, 1998.
Carroll, Jim, Rick Broadhead, and Don Cassell. *Canadian Internet Handbook: 1996 Edition*. Scarborough, ON: Prentice Hall, 1996.

Cassidy, John. *Dot.Con: How America Lost Its Mind and Money in the Internet Era.* New York: Perennial, 2003.

Cavender, Sasha. "On-Line World Miffed over Cyberspace 'Day' Project." *San Francisco Examiner*, February 5, 1996.

CERN. "The Birth of the Web." November 12, 2013. http://home.web.cern.ch/about/birth-web.

Ceruzzi, Paul. *A History of Modern Computing.* 2nd ed. Cambridge, MA: MIT Press, 1998.

Chernow, Ron. "A Market in Need of a Broker." *New York Times*, April 16, 2000, sec. 4, 15.

Chun, Wendy H. K. *Programmed Visions: Software and Memory.* Cambridge, MA: MIT Press, 2011.

Cirillo, Rich. "User Advocate or Enemy of Creativity?" *VARBusiness*, February 19, 2001.

Cleland, Kim. "Fear Creates Strange Bedfellows." *Advertising Age*, March 13, 1995, 16–18.

———. "Upstart Company Snares Big Internet Marketing Gigs." *Advertising Age*, December 12, 1994, 24.

Cline, Craig, and Mark Walter. "24 Hours in Cyberspace: Real-World Prototype for a Surreal Market." *Seybold Report on Desktop Publishing* 10, no. 6 (1996): 3.

Clinton, Bill. "Remarks of Governor Bill Clinton at the Wharton School of Business." University of Pennsylvania, Philadelphia, April 16, 1992. Accessed May 6, 2010, http://www.ibiblio.org/nii/econ-posit.html.

Clinton, Bill, and Al Gore. *Putting People First: How We Can All Change America.* New York: Times Books, 1992.

"Club Med Rockets along the Information Highway via *HotWired.*" PR Newswire, October 27, 1994.

"Computers Seen Beating TV as Infopike Route of Choice." *Daily Variety*, March 3, 1994, 10.

Conklin, Jeff. "Hypertext: An Introduction and Survey." *IEEE Computer* 20, no. 9 (1987): 17–41.

Cook, David. "Auteur Cinema and the 'Film Generation' in 1970s Hollywood." In *The New American Cinema*, edited by Jon Lewis, 11–37. Durham, NC: Duke University Press, 1998.

Cook, Richard. "Behind the Hype: How Agencies Run Their New-Media Business." *Campaign*, August 22, 1997, 37.

Cooper, Michael J., Orlin Dimitrov, and P. Raghavendra Rau. "A Rose.Com by Any Other Name." *Journal of Finance* 56, no. 6 (2001): 2371–88.

Copilevitz, Todd. "How Much Is That Web Site in the Windows?" *Dallas Morning News*, October 7, 1996, 1C.

Couldry, Nick, and José van Dijck. "Researching Social Media as If the Social Mattered." *Social Media + Society* 1, no. 2 (2015): 1–7.

Cre@te Online: The Web Designer's Bible, issue 001 (July 2000).

Cross, Nigel. *Design Thinking: Understanding How Designers Think and Work*. New York: Berg, 2011.

Curtin, Michael, and Thomas Streeter. "Media." In *Culture Works: Essays on the Political Economy of Culture*, edited by Richard Maxwell, 225–50. Minneapolis: University of Minnesota Press, 2001.

CyberHound's Internet Guide to the Coolest Stuff Out There. Detroit: Visible Ink, 1995.

D'Angelo, Frank. "Happy Birthday, Digital Advertising!" *Advertising Age*, October 26, 2009. http://adage.com/article/digitalnext/happy-birthday-digital-advertising/139964/.

Daston, Lorraine. "The Coming into Being of Scientific Objects." In *Biographies of Scientific Objects*, edited by Lorraine Daston, 1–14. Chicago: University of Chicago Press, 2000.

Davis, Glenn. "Announce: Cool Site of the Day." Post to Google Groups, August 10, 1994. https://groups.google.com/forum/#!msg/comp.infosystems.www.misc/KxkWnUnhyX8/oGzKCVcdLRwJ.

Davis, Joshua. *Flash to the Core: An Interactive Sketchbook*. Indianapolis: New Riders, 2002.

———. *PrayStation Hardrive*. CD-ROM. Hong Kong: Systems Design, 2001.

———. Q&A following "Creating Complexity from Simplicity." Presentation at the Flash Forward 2000 Conference, New York City, July 24–26, 2000.

Dayan, Daniel, and Elihu Katz. *Media Events: The Live Broadcasting of History*. Cambridge, MA: Harvard University Press, 1994.

Dear, Brian. "Revisiting the Original 1992 *Wired* Media Kit." *Brainstorms* (blog), April 16, 2013. http://brianstorms.com/2013/04/revisiting-the-original-1992-wired-media-kit.html.

Deleuze, Gilles, and Félix Guattari. *A Thousand Plateaus: Capitalism and Schizophrenia*. Minneapolis: University of Minnesota Press, 1987.

DiSalvo, Carl. "Spectacles and Tropes: Speculative Design and Contemporary Food Cultures." *Fibreculture Journal* 20 (2012): 109–22.

Donath, Judith. "The Establishment of Identity in Virtual Communities." Paper presented at the Seminar on People, Computers, and Design, Stanford, CA, November 1995.

———. "Inhabiting the Virtual City: The Design of Social Environments for Electronic Communities." PhD diss., Massachusetts Institute of Technology, 1996.

Donaton, Scott. "Ketchum Communications." *Advertising Age*, August 22, 1994, 17.

Dougherty, Dale. "Why Flash Is Significant." O'Reilly Network, February 2, 2001. http://archive.oreilly.com/pub/a/network/2001/02/02/epstein.html.

Doull, David. Comment on "The Golden Age of Flash," by Owen van Dijk. *Embrace the Timeline* (blog), October 28, 2003. http://ohwhen.wordpress.com/2003/10/28/the-golden-age-of-flash.

Dovey, John, and Helen W. Kennedy. *Game Cultures*. Maidenhead, UK: Open University Press, 2006.

Doyle, T. C. "Masters of the e-Channel: Web Integrators Forge New Business Models as They Lead the Way into e-Commerce." *VARBusiness*, June 7, 1999, 46.

Drucker, Peter F. *Landmarks of Tomorrow: A Report on the New.* New York: Harper, 1959.

Dunne, Anthony, and Fiona Raby. *Speculative Everything: Design, Fiction, and Social Dreaming.* Cambridge, MA: MIT Press, 2013.

Dwiggins, William Addison. *Layout in Advertising.* New York: Harper, 1928.

Edwards, Paul N., Lisa Gitelman, Gabrielle Hecht, Adrian Johns, Brian Larkin, and Neil Safier. "AHR Conversation: Historical Perspectives on the Circulation of Information." *American Historical Review* 116, no. 5 (2011): 1393–1435.

Eglash, Ron. "Appropriating Technology: An Introduction." In *Appropriating Technology: Vernacular Science and Social Power,* edited by Ron Eglash, Jennifer Croissant, Giovanna Di Chiro, and Rayvon Fouche, vii–xxi. Minneapolis: University of Minnesota Press, 2004.

Elmer-Dewitt, Philip. "Take a Trip into the Future on the Electronic Superhighway." *Time,* April 12, 1993, 50–56.

Emmrich, Stuart. "Agencies Prepare Clients for Interactive Videotex Ads." *Advertising Age,* January 16, 1984.

Evenson, Laura. "Wired to Go Online." *San Francisco Chronicle,* October 20, 1994, E1.

Fahey, Mary Jo. *Web Publisher's Design Guide for Windows.* 2nd ed. Scottsdale, AZ: Coriolis Group, 1997.

First International Conference on the World-Wide Web. "WWW94—Awards." CERN, Geneva, Switzerland, May 25–27, 1994. http://www94.web.cern.ch/WWW94/Awards0529.html.

Flichy, Patrice. *The Internet Imaginaire.* Cambridge, MA: MIT Press, 2007.

Foster, Hal, ed. *Vision and Visuality.* Seattle: Bay, 1988.

Foucault, Michel. *The Archaeology of Knowledge.* Translated by Rupert Swyer. New York: Pantheon Books, 1972.

———. "Confessions of the Flesh." In *Power/Knowledge,* edited by Colin Gordon, 194–228. New York: Pantheon Books, 1980.

Frana, Philip L. "Before the Web There Was Gopher." *IEEE Annals of the History of Computing* 26, no. 1 (2004): 20–41.

Frank, Thomas. *One Market under God: Extreme Capitalism, Market Populism, and the End of Economic Democracy.* New York: Doubleday, 2000.

Frankel, Max. "Word & Image: Liftoff." *New York Times,* November 13, 1994. http://www.nytimes.com/1994/11/13/magazine/word-image-liftoff.html.

Freedom Forum Media Studies Center. *Special Report: Separating Fact from Fiction on the Information Superhighway.* New York: Freedom Forum Media Studies Center, April 1994.

Fry Multimedia. "Website Design for Ragu." 2001. http://web.archive.org/web/20011230001536/http://www.fry.com/clients/ragu.asp.

Fuerbringer, Jonathan. "Ruble Throws Another Scare into Markets." *New York Times,* August 14, 1998. http://www.nytimes.com/1998/08/14/business/the-markets-market-place-ruble-throws-another-scare-into-markets.html.

"FutureWave Software Renames CelAnimator to FutureSplash Animator: New Name Reflects Broader Market and Wide Variety of Internet Browser Support." *Business Wire*, August 12, 1996.

Galison, Peter. *Image and Logic: A Material Culture of Microphysics*. Chicago: University of Chicago Press, 1997.

Gannon, Michael. "From Content to Commerce: VCs Flocked to e-Commerce." *Venture Capital Journal*, January 1, 1999.

Gay, Jonathan. "The History of Flash." Adobe, 2000. https://web.archive.org/web/20071026123951/http://www.adobe.com/macromedia/events/john_gay/page02.html.

Georgenes, Chris. "Flash Designer Gets Hooked on Community." *Edge Newsletter* (Adobe), March 2006. http://www.adobe.com/newsletters/edge/march2006/edge.html?lc_append=500&.

Gibbons, Kent. "Marriage Made on Information Highway Annulled." *Washington Times*, February 24, 1994.

Gibson, William. "Cyberspace '90." *Computerworld*, October 15, 1990, 107.

———. *Neuromancer*. New York: Ace Books, 1984.

Gillespie, Tarleton. "The Stories Digital Tools Tell." In *New Media: Theories and Practices of Digitextuality*, edited by John Caldwell and Anna Everett, 107–26. New York: Routledge, 2003.

Gillies, James, and Robert Cailliau. *How the Web Was Born: The Story of the World Wide Web*. Oxford: Oxford University Press, 2000.

Gillmor, Dan. *We the Media: Grassroots Journalism by the People, for the People*. Sebastopol, CA: O'Reilly Media, 2008.

Girard, Monique, and David Stark. "Heterarchies of Value in Manhattan-Based New Media Firms." *Theory, Culture & Society* 20, no. 3 (2003): 77–105.

González-Vélez, Mirerza. "Assessing the Conceptual Use of Social Imagination in Media Research." *Journal of Communication Inquiry* 26, no. 4 (2002): 349–53.

Green, Tom. "The Rise of Flash Video." *Digital Web*, October 9, 2006. http://www.digital-web.com/articles/the_rise_of_flash_video_part_1/.

Greenspan, Alan. *The Age of Turbulence: Adventures in a New World*. New York: Penguin, 2007.

———. "Remarks by Chairman Alan Greenspan: At the Annual Dinner and Francis Boyer Lecture of the American Enterprise Institute for Public Policy Research." Washington, DC, December 5, 1996. http://www.federalreserve.gov/BOARDDOCS/SPEECHES/19961205.htm.

Greenwald, Ted. "*Wired* 01.01: An Oral History." *Wired* iPad app, April 2013.

Grossberg, Lawrence. "On Postmodernism and Articulation: An Interview with Stuart Hall." *Journal of Communication Inquiry* 10, no. 45 (1986): 45–60.

Grossman, Lev. "You—Yes, You—Are *Time*'s Person of the Year." *Time*, December 25, 2006. http://content.time.com/time/magazine/article/0,9171,1570810,00.html.

Grossman, Wendy. "Futurology: The Aspirant at the Edge of the Universe." *Guardian*, October 19, 1995, B4.

Hahn, Harley, and Rick Stout. *The Internet Yellow Pages*. Berkeley, CA: Osborne McGraw-Hill, 1994.

Hall, Stuart. "Encoding and Decoding in the Television Discourse." Stencilled Occasional Paper 7. Birmingham Centre for Contemporary Culture Studies, 1973.

Hall, Stuart, and Doreen Massey. "Interpreting the Crisis." *Soundings: An Interdisciplinary Journal* 44 (2010): 57–71.

Harvey, David. *The Condition of Postmodernity: An Enquiry into the Origins of Cultural Change*. Cambridge, MA: Blackwell, 1989.

Harvey, Fiona. "Survey—Creative Business—Website Usability Consulting." *Financial Times*, November 28, 2000,.

Hecht, Gabrielle. "Rupture Talk in the Nuclear Age: Conjugating Colonial Power in Africa." *Social Studies of Science* 32, nos. 5–6 (2002): 691–728.

Heid, Jim. "A Call to Action: Making Flash Accessible." *Heidsite*, September 22, 2000. http://www.heidsite.com/archives/flashaccess.html.

Heller, Steven. "Reputations: John Plunkett." *Eye* 28 (Summer 1988). http://www.eyemagazine.com/feature/article/reputations-john-plunkett.

Hershey, Robert D. "Another Sell-Off in Technology Sends Nasdaq into a Correction." *New York Times*, March 31, 2000, A1.

Hewitson, Gillian J. *Feminist Economics: Interrogating the Masculinity of Rational Choice Man*. Cheltenham, UK: Edward Elgar, 1999.

Heydebrand, Wolf, and Annalisa Mirón. "Constructing Innovativeness in New-Media Start-Up Firms." *Environment and Planning A* 34 (2002): 1951–84.

"High Stakes in Cyberspace." *Frontline*, PBS, October 31, 1995. Transcript available at http://www.pbs.org/wgbh/pages/frontline/cyberspace/cyberspacetranscript.html.

Holland, Roberta. "Rob Burgess: 'The Web's For Everybody.'" *ZDNet*, October 27, 2000. http://web.archive.org/web/20010417084227/www.zdnet.com/zdnn/stories/news/0,4586,2646098-2,00.html.

Hu, Tung-Hui. "Real Time / Zero Time." *Discourse: Journal for Theoretical Studies in Media and Culture* 34, no. 2 (2013): 1.

"If at First You Don't Succeed, Try a '.Com.'" Transcript. *CBS MarketWatch*, January 20, 1999.

Information Infrastructure Task Force, National Telecommunications and Information Administration. *The National Information Infrastructure: Agenda for Action*. September 1993.

"Internet, The." *Computer Chronicles*. Hosted by Stewart Cheifet, with guest Brendan Kehole. PBS, 1993. http://archive.org/details/episode_1134.

"Internet Week—Developers Want Flash Flexibility." *VNU NET*, April 5, 2004, 20.

Ip, Greg, and E. S. Browning. "Getting Real: What Are Tech Stocks Worth, Now That We Know It Isn't Infinity?" *Wall Street Journal*, April 17, 2000, A1.

Irani, Lilly C., and M. Six Silberman. "Stories We Tell about Labor: Turkopticon and the Trouble with Design." In *Proceedings of the 2016 CHI Conference on Human Factors in Computing Systems*, edited by Jofish Kaye and Allison Druin, 4573–86. New York: Association for Computing Machinery, 2016.

"iXL Consolidates Acquisitions to Create Largest Internet Solutions Company in Los Angeles: Fastest-Growing Solutions Company Dominates Interactive Entertainment on the Digital Coast." *Business Wire*, May 18, 1998.

"iXL Launches New Interactive Web Site in Conjunction with Grand Opening of New National Headquarters." *PR Newswire*, December 4, 1997.

Jameson, Frederic. *Postmodernism, or, The Cultural Logic of Late Capitalism*. Durham, NC: Duke University Press, 1991.

Jamsa, Kris, and Ken Cope. *World Wide Web Directory*. Las Vegas: Jamsa, 1995.

Jankowsi, Tomasz, et al. *The New Masters of Flash*. Chicago: Friends of Ed, 2000.

Jones, Quentin, and Sheizaf Rafaeli. "Time to Split, Virtually: 'Discourse Architecture' and 'Community Building' Create Vibrant Virtual Publics." *Electronic Markets* 10, no. 4 (2000): 214–23.

Josephson, Sheree, Susan B. Barnes, and Mark Lipton, eds. *Visualizing the Web: Evaluating Online Design from a Visual Communication Perspective*. New York: Peter Lang, 2010.

Jovanovic, Franck. "Economic Instruments and Theory in the Construction of Henri Lefevre's Science of the Stock Market." In *Pioneers of Financial Economics*, edited by Geoffrey Poitras, 191–222. Cheltenham, UK: Edward Elgar, 2006.

Kaplan, David. *The Silicon Boys and Their Valley of Dreams*. New York: William Morrow, 1999.

Kaplan, Jerry. *Startup: Silicon Valley Venture Story*. Boston: Houghton Mifflin, 2005.

Kapor, Mitchell, and John Perry Barlow. "Across the Electronic Frontier." Electronic Frontier Foundation, July 10, 1990. http://w2.eff.org/Misc/Publications/John_Perry_Barlow/HTML/eff.html.

Keegan, Paul. "The Digerati! *Wired* Magazine Has Triumphed by Turning Mild-Mannered Computer Nerds into a Super-Desirable Consumer Niche." *New York Times Magazine*, May 21, 1995, 39–42.

———. "Reality Distortion Field." *Upside*. February 1, 1997, 66–74, 108–14.

Kehoe, Brendan P. *Zen and the Art of the Internet: A Beginner's Guide to the Internet*. Englewood Cliffs, NJ: Prentice-Hall, 1992. http://www.gutenberg.org/ebooks/34.

Kellner, Lauren Fisher. "Trade Dress Protection for Computer User Interface 'Look and Feel.'" *University of Chicago Law Review* 61, no. 3 (1994): 1011–36.

Kelly, Kevin. *New Rules for the New Economy: 10 Radical Strategies for a Connected World*. New York: Viking, 1998.

Kelty, Christopher M. *Two Bits: The Cultural Significance of Free Software*. Durham, NC: Duke University Press, 2008.

Kennedy, Helen. *Net Work: Ethics and Values in Web Design*. Basingstoke, UK: Palgrave Macmillan, 2012.

Kimbell, Lucy. "Rethinking Design Thinking: Part I." *Design and Culture* 3, no. 3 (2011): 285–306.

Kindleberger, Charles. *Manias, Panics, and Crashes: A History of Financial Crises*. 4th ed. New York: Wiley, 2000.

King, Geoff. "From Auteurs to Brats: Authorship in New Hollywood." In *New Hollywood Cinema: An Introduction*, 85–115. New York: Columbia University Press, 2002.

Kirschenbaum, Matthew. *Mechanisms: New Media and the Forensic Imagination*. Cambridge, MA: MIT Press, 2012.

Kirsner, Scott. "The Art of the I-deal: It's All about Speed." *Sydney Morning Herald*, May 12, 1999, 30.

Knafo, Samuel. "Financial Crises and the Political Economy of Speculative Bubbles." *Critical Sociology* 39, no. 6 (2013): 851–67.

Krol, Ed. *The Whole Internet User's Guide and Catalog*. Sebastopol, CA: O'Reilly Media, 1992.

Krugman, Paul. "Roller-Coaster Markets." *New York Times*, April 5, 2000, A23.

Lacey, Kate. *Listening Publics: The Politics and Experience of Listening in the Media Age*. Cambridge, UK: Polity, 2013.

Laderman, Jeffrey M., and Geoffrey Smith. "Internet Stocks: What's Their Real Worth?" *Business Week*, December 14, 1998, 120–22.

Landow, George P. *Hypertext 2.0: The Convergence of Contemporary Critical Theory and Technology*. Baltimore: Johns Hopkins University Press, 1992.

Landow, George P., and Paul Delany. "Hypertext, Hypermedia and Literary Studies: State of the Art." In *Hypermedia and Literary Studies*, edited by Paul Delany, and George Landow, 3–50. Cambridge, MA: MIT Press, 1991.

Law, John. *After Method: Mess in Social Science Research*. New York: Routledge, 2004.

Lee, Edmund. "Designer's Republic." *Industry Standard*, May 29, 2000, 138.

Lefèvre, Henri. *Traité des valeurs mobilières et des opérations de Bourse: Placement et spéculation* [Treatise on financial securities and stock exchange operations]. Paris: E. Lachaud, 1870.

Leggett, Richard Michael, Weyert de Boer, and Scott Janousek. *Foundation Flash Applications for Mobile Devices*. Berkeley, CA: Friends of Ed, 2006.

Lemay, Laura. *Teach Yourself Web Publishing with HTML in a Week*. Indianapolis: SAMS, 1995.

Lemonick, Michael D. "The Net's Strange Day." *Time*, February 19, 1996, 55.

Levy, Steven. "How the Propeller Heads Stole the Electronic Future." *New York Times Magazine*, September 24, 1995. http://www.nytimes.com/1995/09/24/magazine/how-the-propeller-heads-stole-the-electronic-future.html.

Lewis, Herschell Gordon. "Copywriting for Interactive Media: New Rules for the New Medium." In *Interactive Marketing: The Future Present*, edited by Edward Forrest and Richard Mizerski, 229–39. Chicago: NTC Business Books, 1996.

Lewis, Jon. *Hollywood v. Hard Core: How the Struggle over Censorship Saved the Modern Film Industry*. New York: NYU Press, 2002.

Lewis, Peter H. "Help Wanted: Wizards on the World Wide Web." *New York Times*, November 6, 1995, D8.

———. "Sneering at a Virtual Lynch Mob." *New York Times*, May 11, 1994. http://www.nytimes.com/1994/05/11/business/business-technology-sneering-at-a-virtual-lynch-mob.html.

Liu, Alan. *The Laws of Cool: Knowledge Work and the Culture of Information*. Chicago: University of Chicago Press, 2004.

Livingstone, Sonia, ed. *Audiences and Publics: When Cultural Engagement Matters for the Public Sphere*. Bristol, UK: Intellect Books, 2005.

Lockwood, Thomas. *Design Thinking: Integrating Innovation, Customer Experience, and Brand Value*. New York: Allworth, 2009.

Lotz, Amanda. "Textual (Im)Possibilities in the U.S. Post-Network Era: Negotiating Production and Promotion Processes on Lifetime's Any Day Now." *Critical Studies in Media Communication* 21, no. 1 (2004): 22–43.

Lowenstein, Roger. *When Genius Failed: The Rise and Fall of Long-Term Capital Management*. New York: Random House, 2000.

Lunenfeld, Peter. "19?? to 20?? The Long First Decade of Web Design." Presentation at the A Decade of Web Design Conference, Amsterdam, Netherlands, January 21, 2005. Transcripts archived July 2, 2007, https://web.archive.org/web/20070702034246/http://www.decadeofwebdesign.org:80/transcribes.html.

Lynch, Stephen. "Flash a Smile Icon for Net Snapshot." *Orange County Register*, October 9, 1995.

Lyons, Daniel. "Lack of Hard Numbers Fails to Deter Stampede to the Web." *InfoWorld*, November 6, 1995, 61.

Lyotard, Jean-François. *The Postmodern Condition: A Report on Knowledge*. Minneapolis: University of Minnesota Press, 1984.

MacGregor, Chris. "Attention Macromedia: I Will Not Be Your Scapegoat." *Flazoom.com*, November 1, 2000. http://www.flazoom.com/news/scapegoat_11012000.shtml.

———. "A Cancer on the Web Called Flash." *Flazoom*, June 1, 2000. http://www.flazoom.com/news/cancer_06012000.shtml.

Mackay, Charles. *Extraordinary Popular Delusions and the Madness of Crowds*. 1841. Reprint, New York: Harmony Books, 1980.

Mackay, Hughie, and Gareth Gillespie. "Extending the Social Shaping of Technology Approach: Ideology and Appropriation." *Social Studies of Science* 22, no. 4 (1992): 685–716.

Macromedia. "Macromedia and Usability Guru Jakob Nielsen Work Together to Improve Web Usability." Press release. June 3, 2002. http://www.adobe.com/macromedia/proom/pr/2002/macromedia_nielsen.html.

———. "Macromedia Flash 5: Design a Site for Usability Contest." August 13, 2001. http://web.archive.org/web/20010813113404/www.macromedia.com/software/flash/special/designasite/.

———. "Macromedia—Flash Usability." January 23, 2001. http://web.archive.org/web/20010123221700/http://www.macromedia.com/software/flash/productinfo/usability.

———. "Macromedia Flash: Usability Examples." August 5, 2001. http://web.archive.org/web/20010805091105/www.macromedia.com/software/flash/productinfo/usability/examples/.

———. "Macromedia Flash: Usability Quotes." June 15, 2001. http://web.archive.org/web/20010615045829/www.macromedia.com/software/flash/productinfo/usability/quotes/.

———. "Macromedia Launches Usability Initiative." Press release. December 6, 2000. http://web.archive.org/web/20010414033133/www.macromedia.com/macromedia/proom/pr/2000/index_usability.fhtml.

———. "Macromedia's Top 10 Usability Tips for Flash Web Sites." January 23, 2001. http://web.archive.org/web/20010123230300/www.macromedia.com/software/flash/productinfo/usability/tips/.

———. "Showcase Submission Guidelines." November 9, 2001, http://web.archive.org/web/20011109082659/www.macromedia.com/showcase/submit/.

Maddox, Kate. "iXL-Los Angeles Continues on Agency Buying Spree: Spin Cycle and Digital Planet Latest Purchase." *Advertising Age*, May 25, 1998, 14.

Madrick, Jeff. *The Business Media and the New Economy*. Cambridge, MA: Joan Shorenstein Center on the Press, Politics and Public Policy, John F. Kennedy School of Government, Harvard University, 2001. http://dev.shorensteincenter.org/wp-content/uploads/2012/03/r24_madrick.pdf.

Mama's Cucina. "Mama's Cookbook." 1997. http://web.archive.org/web/19970101090944/www.eat.com/cookbook/index.html.

Mandel, Michael. "The Triumph of the New Economy." *Business Week*, December 30, 1996, 68–70.

Manovich, Lev. *The Language of New Media*. Cambridge, MA: MIT Press, 2002.

Marchand, Roland. *Advertising the American Dream: Making Way for Modernity, 1920–1940*. Berkeley: University of California Press, 1985.

"Market's Breathtaking Plunge, The." *New York Times*, April 15, 2000, A16.

Markham, Annette, and Nancy Baym. *Internet Inquiry: Conversations about Method*. Los Angeles: Sage, 2009.

Markoff, John. "Building the Electronic Superhighway." *New York Times*, January 24, 1993, C1.

Marrelli, Charles. "Anatomy of a Web Advertisement: A Case Study of Zima.com." Modem Media website, July 1995. Archived November 11, 1996, http://web.archive.org/web/19961111095536/http://www.modemmedia.com:80/clippings/articles/anatomy.html.

Martin, Roger L. *The Design of Business: Why Design Thinking Is the Next Competitive Advantage*. Boston: Harvard Business Press, 2009.

Martin, Teresa, and Glenn Davis. *The Project Cool Guide to HTML*. New York: Wiley, 1997.

Marwick, Alice E. *Status Update: Celebrity, Publicity, and Branding in the Social Media Age*. New Haven, CT: Yale University Press, 2013.

Marwick, Alice E, and danah boyd. "I Tweet Honestly, I Tweet Passionately: Twitter Users, Context Collapse, and the Imagined Audience." *New Media & Society* 13, no. 1 (2011): 114–33.

Matsumoto, Craig. "AT&T Pulls Plug on Ambitious EO Pen-Computing Business." *Business Journal–San Jose*, August 1, 1994, 4.

McAlester, Duncan, and Michelangelo Capraro. *Skip Intro: Flash Usability and Interface Design*. Indianapolis: New Riders, 2002.

McGee, Suzanne, and Robert McGough. "Some Bulls Make a Case for Tech Stocks—Rationale Shifts for Prized Names." *Wall Street Journal*, April 17, 2000, C1.

McLuhan, Marshall, and Quentin Fiore. *The Medium Is the Massage: An Inventory of Effects*. New York: Random House, 1967.

McMillan, Sam. "The Death of Advertising May Be Premature." *Communication Arts*, December 1994, 212–17.

McPherson, Tara. "Reload: Liveness, Mobility, and the Web." In *The Visual Culture Reader*, 2nd ed., edited by Nicholas Mirzoeff, 458–70. New York: Routledge, 2002.

Meeker, Mary. *U.S. and the Americas Investment Research*. Report, Morgan Stanley Dean Witter. September 26, 1997.

Messina, Judith. "Crunch Time in Alley: Pressured by Competition and Clients, New Media Companies Consolidate." *Crain's New York Business*, May 18, 1998.

Miège, Bernard. *The Capitalization of Cultural Production*. New York: International General, 1989.

———. "The Logics at Work in the New Cultural Industries." *Media, Culture & Society* 9, no. 3 (1987): 273–89.

Milburn, Ken, and Janine Warner. *The Flash 2 Web Animation Book*. Research Triangle Park, NC: Ventana, 1997.

Mirzoeff, Nicholas. "On Visuality." *Journal of Visual Culture* 5, no. 1 (2006): 53–79.

———. *The Right to Look: A Counterhistory of Visuality*. Durham, NC: Duke University Press, 2011.

MIT Media Lab. "About the Book." *A Day in the Life of Cyberspace*, 1996. https://web.archive.org/web/19961031093855/http://www.1010.org/bitsoflife/aboutbook.html.

———. "About This Event." *A Day in the Life of Cyberspace*, 1995. http://web.archive.org/web/19961031093131/http://www.1010.org/Dynamo1010.cgi/about.dyn/__$sid=AnOWIfydzKCAAMec4JDWyB&$pid=0.

———. "1010 Revisited." *A Day in the Life of Cyberspace*, 1996. https://web.archive.org/web/19961031093313/http://www.1010.org/bitsoflife/1010.html.

Mok, Clement. *Designing Business: Multiple Media, Multiple Disciplines*. El Paso, TX: Hayden Books, 1996.

———. "Netobjects." Clement Mok's website. Archived August 4, 2016, https://web.archive.org/web/20160804021035/http://clementmok.com/career/company.php?offset=20&CoID=6&.

"Molson Adds Personal E-mail to Its Site." *Interactive Age*, July 31, 1995, 19.

Montfort, Nick, and Ian Bogost. *Racing the Beam: The Atari Video Computer System*. Cambridge, MA: MIT Press, 2009.

Moock, Colin. Foreword to *Flash Hacks*, edited by Sham Bhangal, ix–xi. Beijing, China: O'Reilly, 2004.

Moody, Glyn. "Internet: Learning to Crawl All over the Web." *Guardian*, December 1, 1994, 5.

Morgenson, Gretchen. "Seeing a Fund as Too Big to Fail, New York Fed Assists Its Bailout." *New York Times*, September 24, 1998. http://www.nytimes.com/1998/09/24/

business/the-markets-seeing-a-fund-as-too-big-to-fail-new-york-fed-assists-its-bailout.html.

Morrison, Dianne See. "Focus—Usability Is the Next Challenge for the Net." *Independent*, March 21, 2001, 2.

Mosaic Communications Corporation. "What's New: August 1993." August 1993. http://home.mcom.com/home/whatsnew/whats_new_0893.html.

Na, Gene. "Judge's Remarks." *Communication Arts Interactive Design Annual* 6 (2000). Archived May 4, 2001, https://web.archive.org/web/20010504092826/http://www.commarts.com:80/ca/interactive_d/ca100/jur_gn.html.

Neff, Gina. *Venture Labor: Work and the Burden of Risk in Innovative Industries*. Cambridge, MA: MIT Press, 2012.

Negroponte, Nicholas. *Being Digital*. New York: Knopf Doubleday, 1995.

Nelson, Jonathan. "From Bedroom to Boardroom, the Net Grows Up." *Ad Week Eastern Edition*, November 9, 1998, 160.

Nelson, Theodor H. "Complex Information Processing: A File Structure for the Complex, the Changing and the Indeterminate." In *Association for Computer Machinery: Proceedings of the 20th National Conference*, edited by Lewis Winner, 84–100. Cleveland, OH: ACM, 1965.

Netscape. "Netscape Builds Momentum with Shipment of Netscape Navigator 2.0." Press release. 1996. http://web.archive.org/Web/20020606145103/http://wp.netscape.com/newsref/pr/newsrelease82.html.

Newcomb, Horace, and Robert S. Alley. *The Producer's Medium: Conversations with Creators of American TV*. Oxford: Oxford University Press, 1983.

Newman, Michael Z., and Elana Levine. *Legitimating Television: Media Convergence and Cultural Status*. New York: Routledge, 2012.

"New Poll Confirms Internet as Information Superhighway." *Internet Business News*, November 1, 1994.

Nielsen, Jakob. *Designing Web Usability*. Indianapolis: New Riders, 2000.

———. "The End of Web Design." *Alertbox*. July 23, 2000. http://www.useit.com/alertbox/20000723.html.

———. "Flash: 99% Bad." *Alertbox*, October 29, 2000. http://www.useit.com/alertbox/20001029.html.

———. "Flash: 99% Bad, Update." *UseIt*. June 3, 2002. https://web.archive.org/web/20040213051559/http://www.useit.com/alertbox/20001029.html.

———. "HCI and the Web." Position paper, CHI96 Conference on Human Factors in Computing Systems, Vancouver, April 1996. http://old.sigchi.org/web/chi96workshop/papers/nielsen.html.

Nussbaum, Bruce. "Design Thinking Is a Failed Experiment, So What's Next?" *Fast Company Design*, April 5, 2011. http://www.fastcodesign.com/1663558/design-thinking-is-a-failed-experiment-so-whats-next.

O'Brien, Chris. "High-Tech Slowdown May Be Worsening." *Knight Ridder Tribune Business News*, June 24 2002, 1.

Oelling, Brandon, Michael Krisher, and Ryan Holsten. "The Declination of Independence." *A List Apart*, March 23, 2001. http://www.alistapart.com/articles/declination.

O'Reilly, Tim. "Web 2.0 Compact Definition: Trying Again." *Radar*, December 10, 2006. http://radar.oreilly.com/2006/12/web-20-compact-definition-tryi.html.

Papacharissi, Zizi. "The Virtual Sphere: The Internet as a Public Sphere." *New Media & Society* 4, no. 1 (2002): 9–27.

Parks, Lisa. *Cultures in Orbit: Satellites and the Televisual*. Durham, NC: Duke University Press, 2005.

Parikka, Jussi. *What Is Media Archaeology?* New York: Wiley, 2013.

Paul, Frederic. "Flashing the Web." *MIT's Technology Review*, March 1, 2000.

Pender, Mark. "Analysts See US Stock Plunge as Healthy, Sobering Correction." *Market News International*, April 14, 2000, 45.

Phillips, John. "Agencement/Assemblage." *Theory, Culture & Society* 23, nos. 2–3 (2006): 108–9.

Pinch, Trevor, and Wiebe Bijker. "The Social Construction of Facts and Artifacts: Or How the Sociology of Science and the Sociology of Technology Might Benefit Each Other." *Social Studies of Science* 14 (1984): 399–441.

Plant, Darrel. *Flash 3! Creative Web Animation*. Berkeley, CA: Macromedia, 1998.

Plunkett, John, and Louis Rossetto. *Mind Grenades: Manifestos from the Future*. San Francisco: HardWired, 1996.

Preda, Alex. *Framing Finance: The Boundaries of Markets and Modern Capitalism*. Chicago: University of Chicago Press, 2009.

———. "Informative Prices, Rational Investors: The Emergence of the Random Walk Hypothesis and the Nineteenth-Century 'Science of Financial Investments.'" *History of Political Economy* 36, no. 2 (2004): 351–86.

Proddow, Louise. *Heroes.com: The Names and Faces behind the Dot Com Era*. London: Hodder and Stoughton, 2000.

Puar, Jasbir. "'I Would Rather Be a Cyborg than a Goddess': Becoming-Intersectional in Assemblage Theory." *Philosophia* 2, no. 1 (2012): 49–66.

Rafter, Michelle V. "Keeping the Peace." *Industry Standard*, January 25, 1999.

Ragus, Dack. "Flash Is Evil." *Dack.com*, September 1, 1999. http://www.dack.com/web/flash_evil.html.

Reagan, Ronald. "Remarks and a Question-and-Answer Session with the Students and Faculty at Moscow State University." May 31, 1988. Public Papers of the President, Ronald Reagan Presidential Library. http://www.reagan.utexas.edu/archives/speeches/1988/053188b.htm.

"Re: FLASH: FLASH 99% BAD." *Chinwag* (email discussion list), October 30, 2000. http://chinwag.com/lists/flasher/old-archive/archive-oct-2000/msg01868.shtml.

Reid, Kanaley. "A Day in the Cyberlife: The Net Collects a Diverse Crowd; A Coast-to-Coast Photo Project Will Show Who and How." *Philadelphia Inquirer*, February 5, 1996.

Remington, R. Roger. *American Modernism: Graphic Design, 1920–1960*. New Haven, CT: Yale University Press, 2003.

Renehan, Edward J. *1001 Really Cool Web Sites*. Las Vegas: Jamsa, 1995.

Richardson, Leonard. "The System of the World Wide Web." Crummy (Leonard Richardson's webspace), 2007. http://www.crummy.com/writing/RESTful-Web-Services/system.html.

Richardson, Robert L. *Web.Guide*. San Francisco: Sybex, 1995.

Rifkin, Glenn. "Arguing over the Dawn of Cyberspace's 'Day' Book." *New York Times*, February 5, 1996. http://www.nytimes.com/1996/02/05/business/arguing-over-the-dawn-of-cyberspace-s-day-book.html.

Roback, Missy. "FutureSplash Animator Makes Impressive Debut." *MacWeek*, October 28, 1996, 37.

Rosenberg, Scott. *Say Everything: How Blogging Began, What It's Becoming, and Why It Matters*. New York: Crown, 2009.

Ross, Andrew. *No-Collar: The Humane Workplace and Its Hidden Costs*. Philadelphia: Temple University Press, 2004.

Rossetto, Louis. "The Story of *Wired*." Paper presented at the Doors of Perception 1 conference, Amsterdam, RAI Convention Centre, 1993. Transcript available at http://museum.doorsofperception.com/doors1/transcripts/rosset/rosset.html.

Rothstein, Jandos. "*Wired* Magazine in the Nineties: An Interview with John Plunkett." In *Designing Magazines: Inside Periodical Design, Redesign, and Branding*, edited by Jandos Rothsetein, 109–15. New York: Allworth, 2007.

Routledge, Nick. "Organic Online: A View on the Web." *World 3*, January 19, 2008. http://web.archive.org/web/20080119133649/http://scribble.com/world3/meme1/space/organichtml.

Ryan, James. "What's Cool on Line? The E-Mail Basket, Please." *New York Times*, October 7, 1996. http://www.nytimes.com/1996/10/07/business/what-s-cool-on-line-the-e-mail-basket-please.html.

Sacks, Jen, and Philip Maher. "Wall Street Wires Up the New World of Media." *Investment Dealers' Digest: IDD* 60, no. 49 (December 5, 1994): 14.

Salter, Anastasia, and John Murray. *Flash: Building the Interactive Web*. Cambridge, MA: MIT Press, 2014.

Scannell, Paddy. "For-Anyone-as-Someone Structures." *Media, Culture & Society* 22 (2000): 5–24.

Schultz, Paul, and Mir Zaman. "Do the Individuals Closest to Internet Firms Believe They Are Overvalued?" *Journal of Financial Economics* 59 (2001): 347–81.

Schurr, Stephen. "To Explain Web Mania, Pundits Tiptoe through a Hot Metaphor." *Wall Street Journal*, May 18, 1999, B1.

Schwartz, John. "Propeller Head Heaven." *Newsweek*, January 18, 1993, 62.

Scott, Michael. "Suds Marketers Look to Future and Dive into the Internet." *Vancouver Sun*, July 22, 1995.

Shah, R. C., and J. P. Kesan. "The Privatization of the Internet's Backbone Network." *Journal of Broadcasting and Electronic Media* 51, no. 1 (2007): 1–9.

Shepard, Stephen B. "The New Economy: What It Really Means." *BusinessWeek*, November 17, 1997, 38–40.
Shepter, Joe. *Personal Web Sites*. Gloucester, MA: Rockport, 2002.
Shiller, Robert J. *Irrational Exuberance*. Princeton, NJ: Princeton University Press, 2000.
"Shopping on the Interactive Highway: Barry Diller Portrays the Future of QVC Network—It's a Multimedia World." *Advertising Age*, May 17, 1993, 22.
Simon, Bob. "The Dot-Com Kids." TV broadcast. *60 Minutes*, February 15, 2000. http://www.cbsnews.com/news/the-dot-com-kids/.
Simon, Herbert A. *The Sciences of the Artificial*. 3rd ed. Cambridge, MA: MIT Press, 1996.
Slater, Don. "Social Relationships and Identity Online and Offline." In *Handbook of New Media: Social Shaping and Consequences of ICTs*, edited by Leah Lievrouw and Sonia Livingstone, 536–46. London: Sage, 2002.
Smith, Timothy K., and Thomas R. King. "Hard Sell: Madison Avenue, Slow to Grasp Interactivity, Could Be Left Behind." *Wall Street Journal*, December 7, 1993, A1.
Smulyan, Susan. *Selling Radio: The Commercialization of American Broadcasting, 1920–1934*. Washington, DC: Smithsonian Institution Press, 1994.
Solomon, Jolie, Michael Meyer, Rich Thomas, and Charles Fleming. "The Dueling Dudes." *Newsweek*, October 11, 1993, 56.
Spigel, Lynn. "Object Lessons for the Media Home: From Storagewall to Invisible Design." *Public Culture* 24, no. 3 (2012): 535–76.
Stäheli, Urs. *Spectacular Speculation: Thrills, the Economy, and Popular Discourse*. Translated by Eric Savoth. Stanford, CA: Stanford University Press, 2013.
Standage, Tom. "Menu Masters of the Cyberian Beanfeast." *Scotsman*, July 21, 1995, 17.
Stellin, Susan. "Sorting Out a Future after Dot-Com." *New York Times*, July 11, 2001, 1.
Sterling, Bruce. *The Hacker Crackdown: Law and Disorder on the Electronic Frontier*. New York: Bantam, 1992.
Stevenson, Michael. "Rethinking the Participatory Web: A History of Hotwired's 'New Publishing Paradigm,' 1994–1997." *New Media & Society* 18, no. 7 (2014): 1331–46.
Stone, Allucquére Rosanne. "Will the Real Body Please Stand Up? Boundary Stories about Virtual Cultures." In *Cyberspace: First Steps*, edited by Michael Benedikt, 81–113. Cambridge, MA: MIT Press, 1991.
Streeter, Thomas. "The Moment of *Wired*." *Critical Inquiry* 31, no. 4 (2005): 755–79.
———. *The Net Effect: Romanticism, Capitalism, and the Internet*. New York: NYU Press, 2011.
Streitfeld, David. "High-Tech Anxiety Hits Silicon Valley: Innovation Capital Waits for the Bubble to Burst." *Washington Post*, December 25, 1999, A01.
Suchman, Lucy. "Anthropological Relocations and the Limits of Design." *Annual Review of Anthropology* 40, no. 1 (2011): 1–18.
———. "Configuration." In *Inventive Methods: The Happening of the Social*, edited by Celia Lury, and Nina Wakeford, 48–60. New York: Routledge, 2012.
———. *Plans and Situated Actions: The Problem of Human-Machine Communication*. Cambridge: Cambridge University Press, 1987.

Taffler, Richard A., and David A. Tuckett. "A Psychoanalytic Interpretation of Dot.com Stock Valuations." Social Science Research Network, March 1, 2005. http://ssrn.com/abstract=676635.
Taylor, Cathy. "Mama Says." *Ad Week Western Edition*, September 25, 1995, 26.
——. "Ordering Interactive a La Carte." *Ad Week Western Edition*, April 1, 1996, 29.
——. "Recruiting for the Interactive Age." *Ad Week Eastern Edition*, June 18, 1994, S12.
——. "Z Factor." *Brandweek*, February 6, 1995, 14–16.
Taylor, Dave. *Creating Cool Web Pages with HTML*. Foster City, CA: IDG Books, 1995.
Teague, Jason Cranford. "The Backlash against Flash." *Independent*, May 28, 2001.
"Technology Sector Outlook." Narrated by Kitty Pilgrim, performed by Vivek Rao. *Trading Places*, CNN, December 7, 1998.
Terranova, Tiziana. *Network Culture: Politics for the Information Age*. London: Pluto, 2004.
Terry, Nicolas P. "Gui Wars: The Windows Litigation and the Continuing Decline of 'Look and Feel.'" *Arkansas Law Review* 47, no. 93 (1994): 93–158.
"That Mess on Your Web Site." *MIT's Technology Review*, September 1, 1998. https://www.technologyreview.com/s/400262/that-mess-on-your-web-site.
Thompson, John B. *The Media and Modernity: A Social Theory of the Media*. Stanford, CA: Stanford University Press, 1995.
Thompson, Robert J., and Gary Burns. *Making Television: Authorship and the Production Process*. New York: Praeger, 1990.
Thrift, Nigel. "'It's the Romance, Not the Finance, That Makes the Business Worth Pursuing': Disclosing a New Market Culture." *Economy and Society* 30, no. 4 (2001): 412–32.
Treffiletti, Cory R. "Rick Boyce." In *Internet Ad Pioneers*, 7–8. Lexington, KY: CreateSpace Independent Publishing Platform, 2012.
Turkle, Sherry. *Life on the Screen: Identity in the Age of the Internet*. New York: Simon and Schuster, 1995.
Turlington, Shannon R. *Walking the World Wide Web: Your Personal Guide to the Best of the Web*. Chapel Hill, NC: Ventana, 1995.
Turner, Fred. *From Counterculture to Cyberculture: Stewart Brand, the Whole Earth Network, and the Rise of Digital Utopianism*. Chicago: University of Chicago Press, 2006.
Turow, Joseph. *Breaking Up America: Advertisers and the New Media World*. Chicago: University of Chicago Press, 1997.
——. *Niche Envy: Marketing Discrimination in the Digital Age*. Cambridge, MA: MIT Press, 2006.
"24 Hours in Cyberspace." TV broadcast. Hosted by Forest Sawyer. *ABC News Nightline*, February 9, 1996.
Understanding the Internet. Written and narrated by Robert Duncan. Directed by Andrew Cochran. Alexandria: VA: PBS Video, 1995. VHS. Transcript available at http://web.archive.org/web/19961225183110/http://www.pbs.org/uti/utitranscript.html.

"Usability Blues." *Cre@te Online* 9 (February 2001): 73.
van Dijck, José, and Thomas Poell. "Understanding Social Media Logic." *Media and Communication* 1, no. 1 (2013): 2–14.
Waldron, Rick. "The Flash History." *Flash Magazine*, July 31 2002. http://www.flashmagazine.com/413.htm.
Walton, John. "Web of Visions." *BBC News*, December 3, 2001. http://news.bbc.co.uk/2/hi/in_depth/sci_tech/2000/dot_life/1684640.stm.
Wardrip-Fruin, Noah. "Digital Media Archaeology: Interpreting Computational Processes." In *Media Archaeology: Approaches, Applications, and Implications*, edited by Erkki Huhtamo and Jussi Parikka, 302–22. Berkeley: University of California Press, 2011.
Warner, Michael. *Publics and Counterpublics*. New York: Zone Books, 2002.
Weingarten, Mark. "Flash Backlash." *Industry Standard*, March 5, 2001.
Weise, Elizabeth Reba. "'Day in the Life of' Crew Takes on Cyberspace." Associated Press, February 6, 1996.
Weltevrede, Esther, Anne Helmond, and Carolin Gerlitz. "The Politics of Real-Time: A Device Perspective on Social Media Platforms and Search Engines." *Theory, Culture & Society* 31, no. 6 (2014): 125–50.
"What Is Coolium?" Cool Site of the Day, December 26, 1996. http://web.archive.org/web/19961226191645/http://cool.infi.net/faq.html.
Williams, Raymond. *Marxism and Literature*. Oxford: Oxford University Press, 1977.
Williamson, Debra Aho. "The Latest Pitch: Interactive AOR." *Advertising Age*, January 30, 1995, 13.
Wilson, Nick. "All-Flash a Fast Track to Failure." *SitePoint*, September 26, 2002. http://articles.sitepoint.com/article/flash-fast-track-failure/2.
"Window on the New Economy." Narrated by Charles Molineaux, produced by Laura Rowley. *Business Unusual*, CNN, July 17, 2000.
Wines, Michael. "Clinton Focusing on the World Economy." *New York Times*, July 6, 1994. http://www.nytimes.com/1994/07/06/world/clinton-focusing-on-the-world-economy.html.
Winner, Langdon. "Do Artifacts Have Politics?" *Daedalus* 109, no. 1 (Winter 1980): 121–36.
Wise, J. Macgregor. "Assemblage." In *Gilles Deleuze: Key Concepts*, edited by Charles J. Stivale, 77–87. Montreal: McGill-Queen's University Press, 2005.
Wolf, Gary. "Justin in the Zone." Post to WELL forum, September 10, 1994. http://www.well.com/~hlr/jam/justin/justingary.html.
———. "The (Second Phase of the) Revolution Has Begun." *Wired*, October 1, 1994, 120–21. https://www.wired.com/1994/10/mosaic/.
———. *Wired: A Romance*. New York: Random House, 2003.
Wolff, Michael. *Burn Rate: How I Survived the Gold Rush Years on the Internet*. New York: Simon and Schuster, 1998.
Wolff, Michael, Peter Rutten, Albert F. Bayers, and Kelly Maloni. *Netguide: Your Map to the Services, Information and Entertainment on the Electronic Highway*. New York: Random House, 1994.

Woolgar, Steve. "Technologies as Cultural Artifacts." In *Information and Communication Technologies*, edited by William H. Dutton, 87–102. Oxford: Oxford University Press, 1996.

WWW Virtual Library. "The WWW Virtual Library: History of the Virtual Library." Accessed January 8, 2017, http://vlib.org/admin/history.

Yavovich, B. G. "Burnett's Interactive Crystal Ball." *Advertising Age*, January 24, 1994, 16.

Zeldman, Jeffrey. "Style vs. Design." Adobe Motion Design Center, 2000. http://www.adobe.com/designcenter/dialogbox/stylevsdesign/.

———. "Where Have All the Designers Gone?" Adobe, March 20, 2000. http://web.archive.org/Web/20001006173913/www.adobe.com/Web/columns/zeldman/20000320/main.html.

INDEX

Page numbers in italic refer to figures

1-800-COLLECT, 78–79
2Advanced Studios, 145
9/11, 97, 162, 187, 194
24 Hours in Cyberspace, 22–23, 96, 99, 102, 104–5, 108–9, 114–17, *197*
1001 Really Cool Web Sites, 67, 71–72, 88

Abbate, Janet, 25
ABC Nightline, 114
ABC World News Tonight, 46
accessibility, 6, 18–20, 119, 151, 156, 172–73, 177, 185–86
Adaptive Path, 164, 189–94
Adigard, Erik, 53–*54*
Adobe, 220n66; Illustrator, 143; purchase of Macromedia 220n66
advertising, 50–53, 118, 121–22, 135, 137, 140, 162, 165, 203; banner ads, 78, 80, 83, 145, 182; indirect, 81; interactive, 19, 75–94, 145. *See also* brochureware
Against All Odds, 102. See also *24 Hours in Cyberspace*
A. G. Edwards, 160
agencement, 9, 198. *See also* assemblage
algorithms, 5, 20, 97, 102, 112–13, 142, 220n68, 221n75
Allen, Matthew, 3, 100
Amazon, 133–34, 161, 168–69, 200
American Association of Advertising Agencies (the 4As), 76
American Enterprise Institute, 128
America OnLine, 81, 92

Andersen Consulting, 138
Andreessen, Marc, 39–40, 57
Andrikipoulos, Shaun, 220n42
Anklesaria, Farhad, 29
Antiochia, Antonio, 216n102
Appadurai, Arjun, 11, 14, 55
Apple, 11, 66, 117, 200, 214n25; *Apple v. Samsung*, 66; *Apple Computer, Inc. v. Microsoft Corp.*, 214n25; iPhone, 66; Mac, 40, 50, 181–82; Powerbook, 178
apps, 5–6, 196, 200, 203; killer apps, 41
Arah, Tom, 186
ARPANET, 25–26
Art Technology Group (ATG), *111*, 113, 118–99, *197*
Artzt, Ed, 76–78
assemblage, 7–9, 118, 162, 193–94, 196, 198, 203; method assemblage, 102–5
asynchronous transfer mode (ATM), 106
AT&T, *80–81*, 87, 182
Auerbach, Jon, 105
Augmented Human Intellect Research Center (Augmentation Research Center), 33
Autodesk: Cyberspace Developer's Kit, 48, 213n75

Badboy Records, 158
Barlow, John Perry, 14, 47–49; *Declaration of the Independence of Cyberspace*, 96–99

245

Barneys New York, 151–54, 156, 172
Bear Stearns, 45
Becker, Lane, 190
Behlendorf, Brian, 79
Bell, Daniel, 125
Bell Atlantic, 44–45
Benjamin's Home Page, 73
Berners-Lee, Tim, 3, 5, 25–26, 30–34, 36–39, 56, 91, 101, 195–96
Bezos, Jeff, 132
Bijker, Wiebe, 222n17
Bina, Eric, 39, 57
blogs, 2–3, 6, 103, 120, 167
Bloxham, Mike, 171
bookmarks, 2, 57–58, 186, 211n27. *See also* hotlists
Bourdieu, Pierre, 18, 61
BoxTop, 136, 139–40
Brand, Stewart, 99
Brenner, Robert, 131–32
broadcast media, 81, 102, 104, 108, 114–15, 122, 136, 159, 197
brochureware, 6, 19, 82, 87, 94, 216n80
browsers, 2, 25, 32, 39–42, 46, 53, 55–58, 62, 72, 92, 141, 171, 195, 202; Browser Wars, 141–43, 189. *See also individual browsers*
Brunsdon, Charlotte, 61
BT Alex Brown, 220n42
Buchanan, Richard, 12
Buffett, Warren, 161
bulletin board systems (BBS), 47, 77, 99, 172
Burgess, Rob, 174–75
Burstein, Daniel, 46–47
Bush, Vannevar, 34

Cambridge Technology Partners, 138
Campus-Wide Information System (CWIS), 29
Canter, Laurence, 77, 94
Capraro, Michelangelo, 179–80
Cassidy, John, 132, 160
Cavender, Sasha, 108
CBS MarketWatch, 132
CD-ROMs, 21, 39, 77, 83, 115, 136, 142, 154
Ceruzzi, Paul, 43
chat rooms, 72, 92–93, 130, 161
Cheifet, Stewart, 35
Chiat/Day, 80
Chrome, 39
Cisco, 161
class, 10, 13, 16–17, 51, 61, 201–2
Clement Mok Designs, Inc., 117
Clinton, Bill, 31, 42–43, 116, 126–27, 131
Club Med, 78–79, 82
CNBC, 132–33
CNN, 132, 156–58
Cold War, 25, 38, 127
Combs, Sean "Puff Daddy," 156, 158
CommerceWAVE, 137
Communication Arts, 20–21, 76; Interactive Design Awards, 150
Communications Decency Act, 116
Compact Disc Connection, 35
CompuServe, 35, 81, 92
Computer Chronicles, 21; "The Internet," 35
conjuncture, 13, 22, 27–8, 54–5, 64–5.
cool, 14–15, 22, 54, 56–57, 61–72, 74–75, 82–95; cool Flash, 150, 171–76, 188
Cool Site of the Day (CSotD), 14, 63, 65, 67–74, 150
Coors Brewing Company: Zima site, 68, 83–87, 91–92, 216n88
Couric, Katie: "What Is the Internet, Anyway?," 1–2
Cox Cable, 44
Creating Cool Web Pages with HTML, 67
Cre@te Online, 20, 168, 181
cultural studies, 61, 165, 200, 202
Cunniff, Tom, 87–88
Curry, Adam, 31
Curtis, Hillman, 145, 179
Cybergrrl, 68

CyberHound's Internet Guide to the Coolest Stuff Out There, 67, 71, 74
CyberSight, 80, 92
cyberspace, 14, 19, 28, 42, 47–48, 47–49, 55, 57, 64–65, 75, 78, 96–100, 106–7, 122

Danbury Printing and Litho, 53
D'Angelo, Frank, 78
Davis, Glenn, 65, 68, 71
Davis, Jim, 86
Davis, Joshua, 157–58, 170, 172–74, 184; One-Upon-A-Forest, 154–55; PrayStation, 154, 168
Dawes, Brenan, 168
A Day in the Life of Cyberspace (1010), 102, 104–13, 117–18, 197
Delany, Paul, 37
Deleuze, Gilles, 8–9, 103, 198
Department of Defense: Advance Research Projects Agency (ARPA), 25
design, definition of, 10; as critical design thinking, 11–12, 199; as design philosophy, 5–13, 190–92, 198–201; as rational or emotional, 10, 17–18, 64, 123–24, 150, 163, 179, 192–94
Digital Planet, 136, 139–40
DiSalvo, Carl, 12
Disney, 142, 146
dispositif, 124
Donath, Judith, 106, 108, 110
dot-com bubble, 7, 9, 12, 19–20, 201; aftermath of, 6, 23–24, 162–94, 203; boom stage, 6, 21–23, 62, 96–120, 125, 127–35, 160, 162; bust phase, 3, 6, 21, 23, 62, 98, 119, 160, 162–65, 170, 185, 189, 191–92, 194, 203, 222n13; displacement stage, 6, 21–22, 25–95, 130; euphoria stage, 6, 21, 23, 118, 121–59, 162; expansion stage, 6, 21
Dreamless, 174, 222n50
Dreamweaver, 146
Drucker, Peter, 125

Dunne, Anthony, 12
Dwiggins, William Addison, 10
Dynamo, 113, 118

economic crash. *See* financial crises
e-commerce, 6–7, 13, 19, 23, 103, 118–19, 121–59, 161–63, 187, 216n80
Eglash, Ron, 165
Electric Word, 50–51
Electronic Data Systems, 138
electronic frontier, 14, 29, 47, 49, 77, 99
Electronic Frontier Foundation (EFF), 48, 116
Ellis, Bert, 136–37, 139
email, 1–2, 7, 68, 70, 77, 86, 106, 108, 149, 170, 172
Englebart, Doug, 33
Epstein, Bruce, 145
Ernst & Young, 140
E-Trade, 130
European Organization for Nuclear Research (CERN), 30, 32, 56, 73, 91, 174, 195

Facebook, 2–5, 100, 113
Federal Open Market Committee (FOMC), 128, 131
Federal Reserve, 23, 128, 131; "Greenspan put" option, 132. *See also* Greenspan, Alan
feminism, 18, 165, 199
file transfer protocol (FTP), 31, 35–36
financial crises, 3, 23, 51, 131; 1987 economic crash, 51, 221n5; 1998 Asian financial crisis, 23, 131; 2000 economic crash, 160-2; South Sea Company bubble, 15, 21, 130. 2008 economic crash, 3. *See also* dot-com bubble
Fiore, Quentin, 53
Firefox, 39
First International Conference of the World Wide Web: Best of the Web Awards, 57

Flash, 6, 123, 162–63, 192–94, 221n75, 223n58; ActionScript, 146, 184–85, 221n75; *Flash: 99% Good*, 179; Flash Forward Conference and Film Festival, 146, 158; Flex, 23, 185, 187; intros, 13, 19, 164, 168–71, 173, 179–80, 184, 190; MX, 23–24, 164, 185–86; MX Professional, 185; usability campaign, 165–89, 223n80; visuality and, 23, 124, 141–59, 171, 184
Flickr, 187, 194
Foote, Cone, and Belding, 83
Foster, Hal, 66
Foucault, Michel, 8, 124
Fox, 142
Frana, Philip, 32, 36–37
Frank, Thomas, 38
Fraser, Janice, 190
Freehand, 143
Free Range Media, 80
Free Software, 15; Free Software Foundation, 32
Freud, Sigmund, 37
Frontline, 85
Fry Multimedia, 87, 89, 216n102
FutureWave Software: CelAnimator, 182; FutureSplash Animator, 141–42, 182–83; *See also* Flash

Garrett, Jesse James, 190, 192
Gay, Jonathan, 182
geek culture, 14–15, 51, 53, 79, 164
gender, 10, 13, 18, 23, 47, 49, 199–200. *See also* feminism
General Public License (GPL), 32
GeoCities, 4
Gerlitz, Carolin, 112
Gibson, William, 47–49, 213n75
gifs, 13, 73, 220n68
Gillespie, Gareth, 165
Gillespie, Tarleton, 167
Girard, Monique, 99–100
GO Corporation: PenPoint, 182

Goldie, Peter, 177
Goldman Sachs, 45
Good Morning America, 31
Google, 200
Gopher, 22, 26, 28–32, 35–37, 41, 211n27
Gore, Al, 31, 42–43, 127
Gore, Tipper, 96, 99
graphical web, 17, 32, 41, 78
graphic design, 6, 10, 37, 53, 145, 147, 192, 197–98
Greenspan, Alan, 23, 131–32; "irrational exuberance" phrase, 128
Greer, Josh, 140
Grimes, Richard, 71
GTE, 44
Guattari, Félix, 8–9, 103
Gulf War, 51
Gullichsen, Eric, 213n75
Gumbel, Bryant: "What Is the Internet, Anyway?," 1–2
Gyford, Phil, 216n88

Hall, Justin: "Justin in the Zone," 69–70; Justin's Links to the Underground, 70
Hall, Stuart, 27, 38, 200
Harris Heidelberg Press, 53
Harvey, David, 38
Hecht, Gabrielle, 28
Helmond, Anne, 112
Hendricks, Lisa Jazen, 139–40
Hewlett-Packard, 66, 214n25
High-Performance Computing Act, 42
Hillman Curtis (studio), 145. *See also* Curtis, Hillman
Hi ReS!, 145
Holsten, Ryan, 173
Hoover, Herbert, 81
hotlists, 57–63, 67, 71–72, 74. *See also* bookmarks
H's Home Page, 68
HTML (hypertext markup language), 9, 39, 61–63, 65, 67, 72–73, 117, 141–43, 145,

149, 197, 216n80, 220n68; DHTML, 6, 142, 173
HTTP (Hypertext Transfer Protocol), 40
Hughes, Kevin, 59
human-computer interaction (HCI), 10–11, 18, 36, 153, 163, 192–93
HyperCard, 36
hyperlinks, 13, 26, 29, 33, 39, 41–42, 56, 68, 85
hypertext, 32–33, 32–34, 36–37, 56, 60, 62, 73, 86, 113, 169, 172, 196. *See also* HTML

IBM, 40; IBM Global Services, 138
Iger, Bob, 46
image maps, 6, 84, 92–93
Industry Standard, 157
InfiNet, 68
Infobot Hotlist Database, 74
information architecture, 6, 13, 74–75, 137, 156, 164, 169
information superhighway, 22, 29, 41–48, 53, 55, 75–77
Information Technology Association of America, 187
initial public offering (IPO), 21–22, 118, 132, 137, 161
interactivity, 20, 23, 42, 53–55, 66, 68, 108, 112, 135, 136–38, 189; Flash and, 140–57, 162–63, 168–69, 173, 179, 183; interaction designers, 5; interactive advertising, 19, 75–94, 145; interactive future, 22, 26, 28, 43–49, 64, 91, 95, 194, 197
International Style, 10
Internet Archive, 20
Internet Explorer (IE), 39, 141–43, 189
internet service providers (ISP), 68, 92
Irani, Lily, 201
iShopHere, 178
iXL, 136–40, 190

Jackass, 174
Jameson, Fredric, 38

Javanovic, Franck, 17
JavaScript, 123, 173
Jerry's Guide to the World Wide Web/Fast Track to Mosaic. *See* Yahoo!
Juxt Interactive, 145, 167
J. Walter Thompson, 80, 87

K2Design, 80
Kahle, Brewster, 20
Kang, Peter, 156–57
Kaplan, Philip: fuckedcompany.com, 187
Kapor, Mitchell, 14, 48–49
Keegan, Paul, 52
Kehoe, Brendan, 35
Kelty, Chris, 14–15
Kennedy, Helen, 209n13
Ketchum Interactive, 77
Kimball, Lucy, 11
Kindleberger, Charles: Minksy-Kindleberger model, 21–23, 27, 130
King, Larry, 44
Kioken, 145, 151–52, 154, 156–58
Kline, David, 46–47
Kodak, 107
Krisher, Michael, 173
Krol, Ed, 31
Krulwich, Robert, 85
Kuhr, Barbara, 51–52
Kuniavsky, Mike, 190, 193

Landow, George, 37
law, 77, 96–97, 116; antimonopoly laws, 44; copyright law, 47, 89–90; deregulation, 27, 44; intellectual property law, 66; regulation, 42–44, 76, 124–25, 131; trademark law, 66. *See also individual cases and statutes*
Law, John, 103
Lefèvre de Châteaudun, Henri, 16–17
Lemay, Laura, 61–62
Levine, Elana, 18
Levy, Steven, 46
Lewinsky, Monica, 131

Lewis, Peter, 77
Library of Congress: Vatican Exhibit, 60
A List Apart, 173
Liu, Alan, 36, 63–64, 68, 166
Long-Term Capital Management (LTCM), 131
Lopez, Jennifer, 156
Lotus, 48
Lotz, Amanda, 82
Lunenfeld, Peter, 147
Lynch, Kevin, 185
Lyotard, Jean-François, 37

MacGregor, Chris, 181; Flazoom, 174–75
Mackay, Charles, 15
Mackay, Hughie, 165
MacLaren McCann, 92, 94
Macromedia, 23–24, 141, 143, 145, 150, 157, 164; acquisition by Adobe, 220n66; Design for Usability contest, 178; Director, 142, 146; Shockwave, 142, 148, 173; usability awareness campaign, 165–89, 223n80. *See also* Flash
Macworld Expo, 53
Mandel, Michael, 129
Manovich, Lev, 182
Marchand, Roland, 81
MarchFirst, 190
marketing, 76, 107, 146, 159, 165–66, 181, 189, 201; branding, 24, 52, 80, 82–94, 133–35, 138, 145, 156–58, 164, 177, 192; content marketing, 82–94; content syndication, 121; data-driven, 19; direct sales, 83, 121, 135; partnerships, 121; relationship marketing, 83, 118; sponsorship, 77–78, 91–92, 121; stickiness, 145. *See also* advertising
Marrelli, Charles, 84–86
Martin, Teresa, 65
Marxism, 37–38, 201
Massey, Doreen, 27
McAlester, Duncan, 179–80
McCahill, Mark, 29, 31

McCann-Erickson Worldwide, 87
MCI, 78–79
McLuhan, Marshall, 53
McPherson, Tara, 70
Mechanical Turk, 201
media studies, 61, 102, 199–201
Meeker, Mary, 133
Mendoza, Gabo: Gabocorp, 143–44, 170
mergers and acquisitions, 23, 44, 119, 122, 135–40, 142, 162, 189–90
Merholz, Peter, 190, 192
Messner Vetere Berger McNamee Schmetterer (MVBMS), 78–80
Metcalfe, Jane, 50–51, 53
methodology, of book, 18–21, 201–3
Microsoft, 66, 142, 214n25; Internet Explorer (IE), 39, 141–43, 189; Office, 166; Windows, 40, 84, 143, 181–82, 214n25
Miège, Bernard, 140, 154, 159
Minsky, Hyman, 21; Minksy-Kindleberger model, 21–23, 27, 130
Mirzoeff, Nicholas, 66
MIT Media Lab: *A Day in the Life of Cyberspace* (*1010*), 102, 104–13, 117–18, 197
Modem Media, 80, 83, 85–87, 135, 190
Modem Media/Poppy Tyson, 190
Molineaux, Charles, 157
Molson Breweries: I Am Online, 91–94
Mondo2000, 51
Moock, Colin, 145
Mosaic, 22, 39–42, 46, 53, 56–61, 69–70, 72–73, 83
MSN, 142
MTV, 31; MTV.com, 31
MySpace, 3

Na, Gene, 156–57
Nakamura, Yugo: Mono*Crafts, 148–49
Napster, 194
NASA, 60, 68, 114
NASDAQ, 4, 6, 22, 160–62, 168, 187, 201

National Center for Supercomputer Applications (NCSA): Mosaic, 22, 39–42, 46, 53, 56–61, 69–70, 72–73, 83
National Information Infrastructure (NII), 31, 127; *National Information Infrastructure: Agenda for Action*, 43
National Radio Astronomy Observatory, 57
National Research and Education Network (NREN), 42
National Science Foundation (NSF), 25, 216n80
navigation schemes, 13, 19, 30–31, 62, 70, 72, 84, 86, 117, 148, 150–52, 154, 156, 177–79, 182
NBC, 1–2. *See also* CNBC
Neff, Gina, 8
Negroponte, Nicholas, 107
Nelson, Jonathan, 79
Nelson, Matthew, 79
Nelson, Ted, 33–34, 37
neoliberalism, 38–39, 44, 165; privatization, 19, 22, 31, 76
Net Café, 21
Netherlands: tulip mania bubble, 15, 130, 133
netiquette, 76, 94, 216n80
.Net magazine, 20
NetObjects: Fusion, 23, 117–19; SitePublisher, 117, 119
Netscape, 22, 58, 92, 109, 141–43, 189
Netscape Communications, 109
network forums, 52, 99–100
New Communalism, 99
New Economy, 4, 7, 23, 100, 121–59, 161, 163, 188, 191, 202
Newman, Michael, 18
new media, 5, 69, 75–78, 91, 121–22, 124, 135, 140–41, 146, 156, 159, 188, 202–3; new media imagination, 19, 22, 25–55, 67, 95
New Visual Economy, 144, 153–54, 158–59, 163

New York Federal Reserve Bank, 131
Nielsen, Jakob, 58, 180–81, 192; "Flash: 99% Bad," 167–70, 172, 174, 185–86; "Jakob's Law of Internet User Experience," 169
Nielsen Norman Group, 223n80
North American Free Trade Agreement (NAFTA), 35
Nussey, Bill, 139
Nynex, 44

O'Connell, G. M., 85
Oelling, Brandon, 173
Office of Scientific Research and Development, 34
Ogilvy, 80
online forums, 14, 48, 68, 72, 170, 173–74
oN-Line System (NLS), 33
OnRamp, 80
open architecture, 26, 37
operating systems, 2, 33, 38, 143, 182, 214n25
Oracle, 119, 161
O'Reilly Media, 98, 101, 103, 199
Organic Online, 79
Our World, 114

Parikka, Jussi, 98
Parks, Lisa, 114
participatory culture, 4, 22, 61, 65, 74–75, 82, 86, 95, 100–102, 107, 120, 200
PathFinder, 77
PBS, 85
Pets.com, 4, 161
Phillips, John, 9
Pinch, Trevor, 222n17
platforms, 2, 4–5, 19, 24, 31, 82, 97–98, 100–102, 105, 107, 112, 119, 125, 143, 164, 185–87, 194, 196–97
Plunkett, John, 51–53
Poell, Thomas, 113
Polaroid, 3

Poppe Tyson, 135, 190
portals, 6, 86, 96, 134, 149–51
postmodernism, 37–38
Preda, Alex, 15, 17, 210n38
print media, 19, 39, 82, 102, 104, 117, 122, 141, 197
Procter & Gamble, 76
Proddow, Louise, 191
Prodigy, 35, 81
Psycho Studio, 168
publics, 15–17, 101–13, 117
Purgusen, Todd, 167

QuickTime, 82

Raby, Fiona, 12
race, 10, 51, 96
radio, 5, 27, 57, 76, 81, 86, 115, 122, 136
Rae Technologies, 117
Ragú: "As the Lasagna Bakes," 88; Mama's Cucina, 87–92; "One Love, One Linguini," 89
Razorfish, 188, 222n13
read-only, 4, 95, 100–101, 103–4, 197
Reagan, Ronald, 126
Regnault, Jules, 16
Remedi Project, 148, 167, 179
Renehan, Edward, 72
Revolution Media, 136
Rheingold, Howard, 47, 50, 70
rich internet applications (RIAs), 6, 24, 164, 185, 187
Rockmore, Alicia, 88
Rolling Stone, 51, 62
Ross, Andrew, 188, 202, 222n13
Rossetto, Louis, 50–51

Safari, 39
Sapient, 138
Scannell, Paddy, 110
Schmetterer, Bob, 78
Scient, 138
search engines, 20, 56, 72, 94

Securities and Exchange Commission (SEC), 21, 131
SGI, 40
Shepard, Stephen, 130
Shepter, Joe, 149
Shiller, Robert, 127
Siegel, Martha, 77, 94
Silicon Alley, 8, 188
Silicon Valley, 3, 98
SitePoint, 176–77
"Skip Intro", 169–*171*, 179, 223n64
Smolan, Rick, 102, 105, 107–9, *115*
Social Construction of Technology (SCOT), 222n17
social media, 2–4, 6, 23, 97–98, 110, 112–13, 196, 199
Sony, 51, 157–58, 174
Southwestern Bell, 44
speculation, 12–14, 54–55, 199; design and, 12–18, 199; 210n28; dot-com bubble and, 5–7, 9, 21–27, 41, 64–65, 123, 128, 130–34, 163, 201; visuality and, 15–18, 66
speculative bubble, 6–7, 9, 12, 17, 21–27, 41–42, 54, 64, 130, 201–202; 210n48
speculative design, 12, 199
Spencer, Herbert, 198
Spigel, Lynn, 10
Spin Cycle Entertainment, 136, 139
Stäheli, Urs, 18
Stallman, Richard, 32
Stanford University, 58–59; Stanford Research Institute, 33
Stark, David, 99–100
startups, 4, 7, 9, 22, 43, 79–80, 109, 117–18, 139–40, 161–62, 189, 199
Sterling, Bruce, 49
Steuer, Jonathan, 79
Stevenson, Michael, 101
stickiness, 145
Stock Exchange, 16, 18, 210n38
stocks and stock market, 4, 15–17, 38, 44, 72, 210n38, 221n5; speculation and, 4–5, 7–8, 15–24, 45, 123–34, 137–38,

159–63, 168, 170, 177, 194, 201–2; valuation models, 4, 133–34, 220n42; volatility and, 18, 123, 160–62; Theoretical Earnings Multiple Analysis (TEMA), 134, 220n42. *See also* dot-com bubble; financial crises
Stone, Allucquére Rosanne, 48
StorySpace, 36
Strategic Internet Services Industry, 137
Streeter, Thomas, 40–41, 47
structures of feeling, 22, 63, 65, 67, 69, 74, 94–95
Suchman, Lucy, 11–12, 104, 201

talent roll-up, 136, 139–40
TCI, 44–45
TCP/IP (Transmission Control Protocol/Internet Protocol), 2, 25–26
technological determinism, 4, 98
Telecommunications Act of 1996, 96
telephone, 2, 30, 42–46, 49, 59, 83, 122, 153, 163, 200. *See also* Apple: iPhone
television, 18, 21, 31, 37, 47–48, 55, 61, 66, 76–77, 80, 82, 114–15, 136, 139, 142, 157, 172, 174, 184, 200, 202–3, 205; cable, 2, 42–46, 49, 75, 121–22, 159, 161; network, 159
telnet, 30–31, 35
Terranova, Tiziana, 37–38
Thai Heritage Project, 68
TheGlobe.com, 132, 161
Thrift, Nigel, 7
Time Warner, 77; Full Service Network, 43–45, 75
Today Show: "What Is the Internet, Anyway?," 1–2
True North: TN Technologies, 87
Turner, Fred, 47, 52, 99
Twitter, 5, 9, 100, 113, 164

UC Museum of Paleontology, 68
Ulm, Josh, 148, 167, 179
Uniform Resource Locator (URL), 7, 20, 36, 86

Uniform Resource Names (URNs), 211n27
University of Cambridge, 71
University of Illinois, Urbana-Champaign, 39
University of Kansas: URouLette, 85–86
University of Maryland Robert H. Smith School of Business: Business Plan Archive, 21
University of Minnesota, 29–32
upgrade culture, 2–3
usability, 5–6, 19, 23–24, 28, 156, 160–94, 223n80
US Congress, 45, 76
UseNet, 31, 77, 90, 94; comp.infosystems. www.misc, 68
useful, 22, 30–31, 40–41, 57–61, 71–75, 81–82, 94, 123–24, 185–86, 194
user experience, 4–5, 18, 24, 71, 118, 150, 160–94, 196
user-friendly, 3, 46–47, 63, 71–72, 117, 123, 176, 196
user-generated content, 3, 5–6, 23, 95, 100–102, 187, 194, 203
US West, 44

Van den Bergh Foods, 87–90
Vanderbilt Television News Archive, 21
van Dijck, José, 113
Vargas, Elizabeth: "What Is the Internet, Anyway?," 1–2
vector graphics, 142–43, 145, 182–83, 220n68
Veen, Jeffrey, 190–91
venture capital, 7–8, 121, 134, 139, 159
versioning, 3, 95, 100–101, 104, 199
Viacom, 44
Viant, 138
video games, 3, 44, 82–84, 87, 146, 154, 162, 185; multiplayer MOOs and MUDs, 72
Videotex, 75–76
Vijn, Yacco: "Skip Intro," 170
virtual reality (VR), 22, 42, 48–49, 53, 55

visuality, 2, 12, 19, 41, 62–63, 67, 71, 73, 92, 108, 114–17, 197–98; "cool" and, 65–67, 90; design and, 9; Flash and, 23, 124, 144–59, 171, 184; futurity and, 46–49; speculation and, 13–18; useability and, 169, 186, 192–93; Web 1.0 and, 4; *Wired* magazine and, 50–54. *See also* New Visual Economy
volumeone, 145

W3C, 195
Walker, Doug, 92, 94
Walking the World Wide Web: Your Personal Guide to the Best of the Web, 60
Wall, Kevin, 140
Wardrip-Fruin, Noah, 98
Warner, Michael, 104, 110–13
WayBack Machine, 20
Web 1.0, 4, 95, 101–3, 198–99, 203. *See also* versioning
Web 1.5, 101, 103
Web 2.0, 3, 6, 95, 98, 100–103, 107, 110, 120, 141, 165, 187, 194, 199, 203
web-authoring software, 5, 21, 24, 119
web design, 4–6, 9–10, 12–20, 41, 100–105, 123, 136–59, 162–73, 193, 201–202
Webby Awards, 21
Web.Guide, 60, 73
WebLouvre, 60
Web service providers, 79–80, 83
web standards, 6, 18, 124, 142, 165, 172–73, 189, 195; Web Standards Project (WaSP), 142
Webvan, 4

Weinman, Lynda, 169
Weltevrede, Esther, 112
Wharton School of Business, 127
Whole Earth Catalog, 52
Whole Earth 'Lectronic Link (WELL), 47–49, 52, 99
Whole Earth Review, 51
wide-area information service (WAIS), 26, 31, 36
Wikipedia, 3, 194
Williams, Raymond, 67, 74
Wired magazine, 22, 42, 50–55, 73, 99; HotWired, 70, 77–81, 83, 101, 190, 193, 216; Wired Ventures, 79
Wolf, Gary, 73–74; "Justin in the Zone," 69–70
Wolff, Michael, 43, 121, 124; Wolff New Media, 121
Woolgar, Steve, 166–67, 222n17
World Wide Web, 2–3, 22, 25–26, 29, 31, 36, 56–60, 88, 91, 106, 157, 195–97
World Wide Web Foundation, 195
World Wide Web Virtual Library, 56

Xanadu, 33
Xerox PARC, 11

Yahoo!, 58–59, 61, 169
Young, Indi, 190
Young & Rubicam, 87
YouTube, 1, 3, 110, 187, 194

ZDNet, 174
Zeldman, Jeffrey, 142, 153

ABOUT THE AUTHOR

Megan Sapnar Ankerson is Assistant Professor of Communication Studies at the University of Michigan. She is co-editor of the international journal *Internet Histories: Digital Technology, Culture and Society*.